creating your own great

dvds & cds

ISBN 0-13-100105-1

9 780131 001053

90000

mark l. chambers
■ creating your own great dvds & cds

tom sheldon
■ upgrading your hp pavilion pc

nancy stevenson
■ hp pavilion pcs made easy

burn it!

creating your own great
dvds & cds

the official hp guide

mark l. chambers

Prentice Hall PTR
Upper Saddle River, NJ 07458
www.phptr.com

A Catalog-in-Publication Data record for this book can be obtained from the Library of Congress.

Editorial/Production Supervision: *Faye Gemmellaro*
Acquisitions Editor: *Jill Harry*
Cover Design Director: *Jerry Votta*
Cover Design: *Talar Boorujy*
Manufacturing Manager: *Alexis R. Heydt-Long*
Manufacturing Buyer: *Maura Zaldivar*
Marketing Manager: *Dan DePasquale*
Editorial Assistant: *Katie Wolf*
Series Design: *Gail Cocker-Bogusz*
Publisher, Hewlett-Packard Books: *Patricia Pekary*

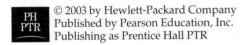 © 2003 by Hewlett-Packard Company
Published by Pearson Education, Inc.
Publishing as Prentice Hall PTR

Printed in the United States of America

10 9 8 7 6 5 4 3 2 1

ISBN 0-13-100105-1

Pearson Education Ltd.
Pearson Education Australia Pty., Ltd.
Pearson Education Singapore, Pte. Ltd.
Pearson Education North Asia Ltd.
Pearson Education Canada, Ltd.
Pearson Educación de Mexico, S.A. de C.V.
Pearson Education—Japan
Pearson Education Malaysia, Pte. Ltd.

This book is dedicated with all a father's love to my daughter,
Chelsea Chambers—"the Chelsea Girl."

Contents

Preface

Who would have ever thought that DVD recording would be so affordable so soon? When I wrote my first book on the subject of optical recording in 1997, a CD recorder was out of budgetary reach for most of us—in fact, I had bought my own personal drive only about three months before! (I had gathered a year or two of previous recording experience, strictly from using a very expensive CD recorder that my company had bought to handle backups at the office.) From the first appearance of CD recorders in the late 1980s, it took a good chunk of a decade for CD recording technology to gain acceptance and hardware and software prices to drop.

Today, CD recording technology is "old hat"—if you're shopping for a computer these days, you'll find that virtually every model has a CD-RW drive as standard equipment. But, unlike the early days of CD recording, it took only a couple of years for affordable DVD recording hardware and software to arrive on the scene. DVD discs can store it all: gigabytes of high-quality digital video, thousands of MP3 audio files, and all of the digital images that a professional photographer can produce in an entire career! (Heck, rewriteable DVD discs are even well suited for mundane chores you've been performing all along, such as storing hard drive backup data.)

In writing this book, I've made a serious—and I hope a successful—attempt at gathering together *all* of the information that a PC owner is likely to need to explore the exciting world of DVD recording. By the way, that includes several chapters that concern CD recording, as well—things such as disc label printing and standard Red Book audio CD recording—just in case you have to return to the "archaic" world of 700 MB from time to time.

Do I Need a Hewlett-Packard Recorder?

Definitely not! Don't get me wrong, I have an HP drive in my own PC—in my personal opinion, they make some of the best hardware on the planet—but like every title in the HP Books series, this book has been expressly written for all PC owners who want to record DVDs and CDs, using any recorder on the market from any manufacturer. In fact, I talk about specific hardware features in only one or two places in the book, and the recommendations and tips I mention will carry over to any recorder.

A Word About Organization

To be honest, this book makes a great linear reading adventure—but only if you're a novice when it comes to optical recording, and you're interested in what makes things tick. If you already have experience recording basic discs, your recorder is already installed, or you're just not interested in how your DVD recorder works, you may decide to skip the material at the beginning and return to it later. This section will help familiarize you with the design of the book.

The first group of three chapters explains how optical recording works—how your computer's CD-ROM or DVD-ROM drive reads files and music from a disc and how your recorder stores information on a blank CD or DVD. You'll also learn how to install an internal or external DVD recorder, using an EIDE, FireWire, or USB connection.

I'll show you how to prepare both your computer and your data before you record, assuring you of top performance, error-free operation, and the best organization for your finished discs.

The next three chapters provide you with complete, step-by-step procedures for burning basic audio and data discs with HP Record-Now, along with drag-and-drop recording within Windows, using HP DLA. I'll cover how you can create your own DVD Video discs for use in most DVD players, using MyDVD.

The final chapters cover the more exotic procedures and subjects in optical recording. You'll learn how to record digital video directly to disc, how to use PowerDVD to watch DVD movies on your PC, and how to print your own professional-looking custom disc labels and jewel box inserts. I'll discuss advanced formats, such as CD-Extra, Video CDs, and multisession discs, and show you how to create each one, step by step. I'll also show you how to create digital photograph slide show discs, how to edit your own digital video movies, and how to archive your existing vinyl albums and cassettes to audio CDs. You'll even learn how to design and produce a powerful menu system for your data discs with the same tools you use to create Web pages! Finally, I provide a software- and hardware-troubleshooting chapter as well.

At the end of the book, you'll find a helpful Glossary (which can aid you in keeping track of what strange term means what), as well as valuable recording and troubleshooting tips from Hewlett-Packard's technical support group. Just in case you haven't bought a recorder yet and you're shopping for one right now, I also provide a quick rundown of the features to look for in a recorder—as well as the "Do's and Don'ts" of buying any computer hardware online.

Watch For Helpful Icons!

Before I launch into Chapter 1, let me familiarize you with the special features you'll find in the text:

• You'll find *Tips* that I've added to help you save time and money—as well as avoid potential pitfalls and recording errors.

 • Pay close attention to any *Caution* icons in the text—whatever it is, you should definitely avoid it!

 • Make sure that you have all of the *Requirements* that I've listed for a project before you begin the step-by-step procedure.

Where to Go Next

Here are my recommendations (don't forget to return later to read chapters that you've skipped):

- If you haven't bought your recorder yet, read through the chapter titled "Tips on Buying Your Recorder"—then begin with Chapter 1.
- If you've already bought your recorder but haven't installed it, begin with Chapter 2.
- If you're interested in mastering the basics of recording and your drive is already working, begin with Chapter 3.
- Finally, if you're already well experienced with burning audio and data CDs and you'd like to jump into recording digital video on a DVD, begin with Chapter 6.

It's my sincere hope that you'll find this book valuable: I hope it answers your questions, provides an occasional chuckle and—most of all—helps you have fun with your DVD recorder. If you have any questions or comments you'd like to send me, please visit my Web site, MLC Books Online, at http://home.mlcbooks.com.

And now…let's *Burn It!*

Acknowledgments

The arrival of this book is a very happy event—my first volume of computer lore for Prentice Hall—and an opportunity to thank several folks who have worked hard together as a team to produce a great book!

First, my appreciation to the great Production staff and editors at Prentice Hall; in particular, Faye Gemmellaro, my Production Editor (who took over the day-to-day schedule for this project at very short notice and did a fantastic job) and Roy Tomalino, my Technical Editor (who scanned every word for accuracy and provided me with cutting-edge information on DVD+R and the latest software)! I'd also like to thank Marti Jones, my Copyeditor, for her attention to detail and hard work in checking my grammar and spelling.

I also owe a great debt of gratitude to Jill Harry, Executive Editor at Prentice Hall PTR, for her tireless efforts during the design of the book and its contents! She coordinated this project from the beginning, and I have her to thank for the chance to work on this new HP title.

Speaking of Hewlett-Packard, I'd like to close these acknowledgements by expressing my heartfelt thanks to three of HP's best: Dean Sanderson, who provided me with technical information and artwork; Walt Bruce, my contact and advisor with HP Books; and finally, Pat Pekary, HP Books Publisher and Program Manager, for both her continued support and her friendship.

About the Author

Mark L. Chambers has been an author, computer consultant, BBS sysop, programmer, and hardware technician for more than 20 years. (In other words, he's been pushing computers and their uses far beyond "normal" performance limits for decades now.) His first love affair with a computer peripheral blossomed in 1984, when he bought his lightning-fast 300BPS modem for his Atari 400—now he spends entirely too much time on the Internet and drinks far too much caffeine-laden soda. His favorite pastimes include collecting gargoyles, St. Louis Cardinals baseball, playing his three pinball machines and the latest computer games, supercharging computers, and rendering 3-D flights of fancy with TrueSpace—and during all that, he listens to just about every type of music imaginable. (For those of his readers who are keeping track, he's up to 1,200+ audio CDs in his collection.)

With a degree in journalism and creative writing from Louisiana State University, Mark took the logical career choice and started programming computers...however, after five years as a COBOL programmer for a hospital system, he decided there must be a better way to earn a living, and he became the documentation manager for Datastorm Technologies, a well-known communications software developer. Somewhere between organizing and writing software manuals, Mark began writing computer books; his first book, *Running a Perfect BBS*, was published in 1994.

Along with writing several books a year and editing whatever his publishers throw at him, Mark has recently branched out into Web-based education, designing, and teaching a number of online classes—called WebClinics—for Hewlett-Packard.

Mark's rapidly expanding list of books includes the *Office v. X for Mac OS X Power User's Guide*, *Building a PC for Dummies*, *Scanners for Dummies*, *CD and DVD Recording for Dummies*, *The Hewlett-Packard Official Printer Handbook*, *The Hewlett-Packard Official Recordable CD Handbook*, *The Hewlett-Packard Official Digital Photography Handbook*, *Computer Gamer's Bible*, *Recordable CD Bible*, *Teach Yourself the iMac Visually*, *Running a Perfect BBS*, *Official Netscape Guide to Web Animation*, and the *Windows 98 Troubleshooting and Optimizing Little Black Book*.

His books have been translated into 12 different languages so far—his favorites are German, Polish, Dutch, and French. Although he can't read them, he enjoys the pictures a great deal.

Mark welcomes all comments and questions about his books—you can reach him at mark@mlcbooks.com.

1

How Does DVD Recording Work?

In This Chapter

✔ Understanding binary language

✔ Reading data from a DVD

✔ Explaining the recording process

✔ Selecting the correct media

✔ Handling and storing your discs properly

✔ Understanding copy-protected discs

✔ Explaining DVD recording formats

Before you jump into installing your DVD recorder and burning your first discs, you should be properly introduced to the world of the digital versatile disc: how information is stored on a DVD and retrieved, what types of media your DVD recorder can use, how to handle your discs properly, and so forth.

I n this chapter, I'll provide you with an overview of what's happening behind the scenes and the basics of DVD storage.

The Language of Computers

Whoa! You may be saying to yourself, "This sounds like we're ramping up to a discussion full of techno-speak." Don't worry, I'll keep things in honest-to-goodness English. In fact, the binary language used by computers is the simplest language ever devised. (Look at the size of this section: It can't be too complex!)

Computers read, write, and communicate to each other using binary, which has only two values: zero and 1, as illustrated in Figure 1.1, just like a light switch. Your computer stores files and programs as long strings of binary data—and the music on an audio compact disc is also recorded in binary (it's the job of an audio CD player to read the digital information on a disc and convert it back to an audio signal).

Using binary, your computer can store audio and video as data files—in fact, you can even store a "snapshot" of the entire contents of a CD or DVD as an image file on your hard drive (provided that you have enough space). You can use this binary image to make copies of a CD or DVD whenever you need them.

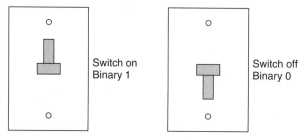

Switch on
Binary 1

Switch off
Binary 0

FIGURE 1.1   There are only two values in the binary language of computers and audio CD players.

Reading Data from a Disc

So how does a DVD (or an audio CD) store binary data? (By the way, DVD actually stands for *digital versatile disc*, although many folks now seem to think it stands for *digital video disc* instead. Just another example of how even the acronyms that techno-types use can change over time.)

Here's why a DVD recorder is technically called an *optical* storage device: the "on" and "off" states of binary information are represented by the presence of (or lack of) laser light. As you can see in Figure 1.2, a commercially made DVD disc is actually composed of several layers. One (or more) of those layers is highly reflective; this aluminum surface is covered by a groove with microscopic *pits*, and these pits are arranged so that they can be read by a laser beam.

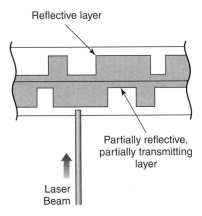

FIGURE 1.2   Although a DVD disc looks like a solid piece of plastic, it's actually more like a sandwich.

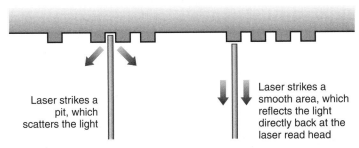

FIGURE 1.3   The basics of optical storage: A laser beam "reads" the pits and lands in a disc.

As you can see in Figure 1.3, when the drive's laser beam hits a pit, it scatters, and most of it isn't reflected back—this represents a zero in binary data. However, when that same laser beam hits one of the flat surfaces (called a *land*), it's reflected cleanly back (much like your reflection in a mirror), and the drive registers that the beam has been returned. This transition between a pit and a land (or, if you remember our light switch, the difference between darkness and light) represents a 1 in the stream of binary data.

Once the drive has read the data, it sends the binary data over a cable connection to your computer. If you're copying a file from a disc, all of those ones and zeroes are saved as a file on your hard drive; if you're running a program directly from the disc or loading a program directly from the disc, the data is piped directly to your computer's processor. Sounds simple, doesn't it? This manipulation of laser light is used by both computer DVD drives and your home DVD player to read the binary data encoded on a disc as pits and lands.

Recording DVDs and Audio CDs

"Now I understand how a DVD disc is read, but how can I *record* one?" As you might have guessed, your DVD recorder must somehow create the equivalent of pits and lands on a blank disc. As you can see in Figure 1.4, a recordable blank disc is composed of layers, as well; one of those layers is *reactive* (meaning its physical properties can be altered).

The reactive layer is made from either a special dye or a special crystalline compound (depending on the type of disc you're recording). When the dye layer is struck by a laser beam, it permanently discolors. When the crystalline layer is hit by a laser beam, it changes, as well. (In the case of rewriteable CDs and DVDs, the crystalline layer can be "reset" later by the same beam and used again.)

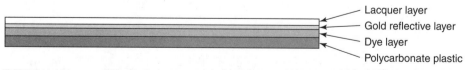

Lacquer layer
Gold reflective layer
Dye layer
Polycarbonate plastic

FIGURE 1.4   A recordable disc contains a layer of dye or a crystalline compound.

The laser beam in a DVD recorder can be set to both low and high power settings; the high power setting is used to create the nonreflective "pit" (Figure 1.5), whereas the low power setting makes no change to the disc, leaving it a land (Figure 1.6).

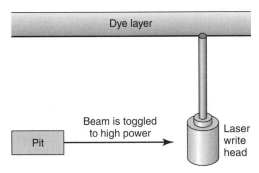

FIGURE 1.5   A high-powered laser beam creates a pit in the dye layer of a recordable disc.

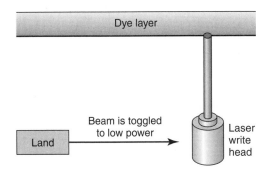

FIGURE 1.6   A low-power laser beam makes no change to the disc, leaving the surface as a land.

When you read a recorded DVD disc, the drive's laser beam strikes the reactive surface, and one of two things happens:

- If the reactive layer has not been changed, the beam is returned by the reflective layer behind it, acting just like a land in a commercially made disc.
- If the reactive layer has been altered, the beam is scattered as it would be by a pit in a commercially made disc.

As I mentioned before, the crystalline layer used by some discs can be altered back and forth between clear and opaque, over and over, so the disc can be completely rewritten; however, these discs must be formatted before they can be reused. This formatting process is very similar to the formatting you've probably performed on a hard drive: It clears the entire surface of the disc, essentially returning the disc to a "factory-fresh" state. (Note that this formatting will begin immediately upon the loading of a DVD+RW disc if you're using an HP DVD-Writer drive—in effect, it happens in the background, so there's no need to worry about formatting.)

Some programs refer to this formatting step as "erasing" the disc; for example, Figure 1.7 illustrates the options you can set in Roxio's Easy CD Creator while erasing a CD-RW disc. (I cover Easy CD Creator in detail later, in Chapter 10, in the section titled "Introducing Easy CD Creator 5.")

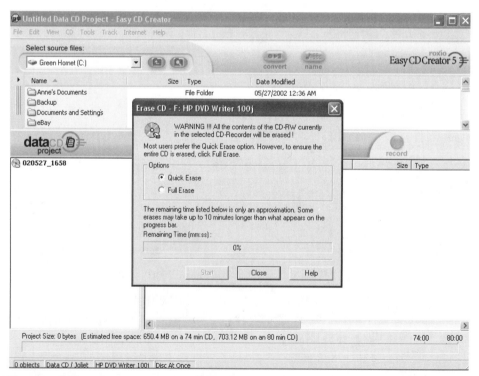

FIGURE 1.7   Formatting a CD-RW disc in Easy CD Creator.

Identifying Your Media

If you've been shopping for optical media lately, you know there's a confusing forest of acronyms and incompatible standards that you must brave before you can click the Buy button—or, if you're in an actual brick-and-mortar store, hand the box to a salesperson. Don't panic, however: In this section, I'll explain the advantages, drawbacks, and compatibilities of each of the major types of CD and DVD media.

CD-R and CD-R/W

CD-R (short for *compact disc recordable*) was the first recordable disc available, and at the time of this writing, it's still the most popular and the least expensive. Currently, a stack of 100 CD-R discs should cost you anywhere from $30 to $50. A typical CD-R disc can store up to 700 MB of computer data or 80 minutes of audio. Because CD-R discs use a dye layer, the data and audio you burn to a CD-R is permanently recorded and can't be erased or formatted.

CD-R discs are the most compatible of CD media because of their high reflectivity. Any CD-ROM drive—no matter how old—can read CD-R discs, and they're the only type of discs that will play in *every* home or car audio CD player.

CD-RW, which stands for *compact disc rewriteable*, is the standard for reusable CD recording; they can store 650 MB of data or 74 minutes of audio. At current prices, a box of 25 CD-RW discs should cost you around $60. As I mentioned earlier in this chapter, you can erase the contents of a CD-RW disc by formatting it, which returns the disc to a blank state, ready to be used again. Unfortunately, the crystalline layer used by a CD-RW disc makes it incompatible with virtually all audio CD players.

DVD+R, DVD-R, DVD+RW, and DVD-RW

DVD+R and DVD-R (predictably, that's short for *DVD recordable*) are often referred to as "write-once" discs. DVD-R is probably the most popular recordable DVD media available at the time of this writing; it was the first to arrive on the market, and it offers the best compatibility with both DVD set-top players and computer DVD-ROM drives.

Both DVD+R and DVD-R have higher compatibility than do rewriteable DVD media because they have higher reflectivity. (However, neither DVD+R nor DVD-R media can be used with a standard CD-ROM drive or audio CD player.) DVD-R discs typically sell for about $20 for a pack of 10 discs, whereas the newer DVD+R discs sell for around $5 each at the time of this writing.

Like a CD-R disc, DVD+R and DVD-R discs can be recorded only once. However, a single-sided disc can hold 4.7 GB, whereas a double-sided disc can hold 9.4 GB of data. A double-sided DVD doesn't have a standard label; printing can appear only around the spindle hole, because a label would interfere with the laser. (You may have already noticed this if you have a DVD player, because many DVD movies have a widescreen version of the film on one side and a standard ratio version on the other side.)

DVD+RW and DVD-RW (which stands for *DVD rewriteable*) are relatively new arrivals on the DVD recording horizon. Many drives capable of writing DVD-R discs also write DVD-RW discs. DVD+RW hit the market slightly later than the DVD-RW products, but it's already making a name for itself among a number of DVD recorder manufacturers and has been chosen by several large PC manufacturers as a component in high-end PCs. Like a CD-RW disc, DVD+RW can be reformatted and reused up to 1,000 times. DVD+RW and DVD-RW discs can store 4.7 GB per side and are compatible with most DVD players and DVD-ROM drives. Naturally, the DVD+RW and DVD-RW discs you record can't be read on a standard CD-ROM drive or audio CD player.

The only drawback to DVD+RW and DVD-RW technology is the price of the media: currently, anywhere from $8 to a whopping $20 per 4.7-GB disc! Rewriteable DVD media (just like its cousin, CD-RW media) is more expensive to manufacture and, therefore, more expensive than DVD+R and DVD-R media. However, the convenience of erasing and rewriting the discs make them well worth the extra expense. As more of these discs are produced, they'll be easier to find, and I predict that their price will drop as rapidly as DVD-R discs have done.

DVD-RAM

DVD-RAM discs are reusable and, like DVD+RW and DVD-RW, they can store up to 9.4 GB of data, using both sides. Like DVD-R, DVD-RAM is a well-established standard, and DVD-RAM discs are popular for system backups and portable mass storage. When you consider the megabyte/dollar ratio, they're significantly cheaper than other types of rewriteable DVD storage, currently averaging about $10 for a double-sided 9.4-GB disc.

However, DVD-RAM is compatible with very few DVD players and no DVD-ROM drives (hence the advantage of the newer DVD+RW and DVD-RW discs), making them a poor choice for creating DVD video discs. A few late-model (also called "fourth-generation") DVD players can read a DVD-RAM disc, as well.

tip **Due to the wide disparity in manufacturers, I can't tell you whether your current DVD player will read a DVD-R, DVD+RW, or DVD-RAM disc—and you may not be able to test its compatibility easily unless you happen to have friends with all three types of drives! Therefore, it's a very good idea to check your DVD player's manual and its specifications for any word on recorded DVD media. Also, you can check the manufacturer's Web site to see whether there's any information on media compatibility.**

Caring for Your Optical Pets

You've probably been handling CDs for many years now—and because DVDs look so similar and work the same way, you might think you can handle and store them in the same way. As you've learned in this chapter, however, DVDs carry a lot more information than do those "antique" compact discs, and they do it by compressing more pits into a smaller area. Therefore, a simple fingerprint, a speck of dust (or, in the worst case, a scratch) can result in read problems that you wouldn't encounter with a CD.

In this section, you'll learn the correct way to handle, clean, and store your DVDs.

DVD Handling 101

Let's begin with the Golden Rule of DVD: *Never allow anything to touch the reflective surface of your discs!* A disc that's smudged by fingerprints, grease, or oil may be unreadable, because the laser beam in your DVD player must be able to pass through the surface of the disc *twice* to read it. Even the cleanest set of fingertips can transfer all sorts of contaminants when you hold a DVD incorrectly.

There are two proper methods of holding a disc: by the outside edge and by inserting a finger into the spindle hole, as shown in Figures 1.8 and 1.9.

Things to Avoid

DVDs have four arch enemies that you should avoid at all costs:

- **Heat.** If you have a car audio CD player, you already know how important it is to protect discs from high temperatures; a slight warping caused by overheating can render a disc unreadable. Therefore, direct sunlight should also be avoided.

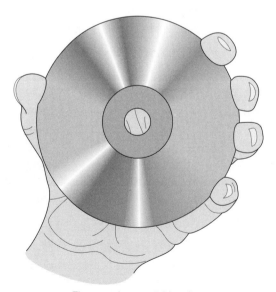

Fingers grip on outside edge

FIGURE 1.8   Holding a DVD by the outside edge.

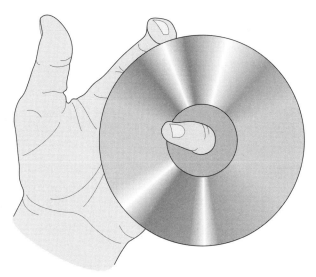

FIGURE 1.9   Hold a disc using the spindle hole (for those with smaller fingers).

- **Sharp objects.** Scratching the surface of a disc can result in read errors (or, in the worst case, ruin the disc if the scratch is deep and wide enough).

tip You can write on the label side of a single-sided DVD disc—as long as the manufacturer says you can print on the surface—but don't use a ballpoint pen! The sharp point of a ballpoint pen can easily damage the disc. Instead, use a quick-drying, soft-tip permanent marker to label your discs. Slow-drying markers may contain solvents that can eat through or etch the outer layer of the disc.

- **Liquids.** Although some disc cleaning systems use liquids (usually based on an alcohol solution), they're not necessary. Other liquids should be wiped off as quickly as possible, using the technique I will demonstrate later in this section.
- **Dust.** Because DVDs store information based on reflected light, dirty discs tend to skip in an audio CD player (or end up unreadable in a DVD player). Keep your discs in their jewel cases or store them properly to keep them free of dust.

tip **Never set a DVD on a table! If you need to set a disc aside—even just for a few seconds—always put it back in its jewel case.**

caution **Do not use a "laser lens cleaner" in your DVD recorder or your DVD player—you could damage your drive! The laser and lens system inside these units doesn't require any cleaning and can be easily damaged by fibers or abrasives.**

Storing Your Discs

Most folks use the jewel box or plastic case that a disc comes with to store it. This is fine, but there are other methods of storing your discs that take up less room. For the best protection for the most discs in the least space, I recommend a DVD disc binder, which stores each disc in an individual plastic sleeve, so you can toss the jewel case. DVD disc binders are available in sizes that range from 10 to more than 250 discs.

You'll also find a bewildering variety of plastic and cloth cases—accordions, clamshells, zippered books, and many more. These "hi-tech" storage units are fine, as long as they're specifically designed for use with DVDs: Never store a DVD in a jewel box or case that's meant for a standard CD. Although the hole in the center of a CD and DVD are the same, a jewel case for a DVD will have a slightly smaller hub diameter (making it easier to remove to avoid flexing). If the disc is flexed, it should be allowed to stabilize and become flat again before trying to read or write to it.

Whatever storage method you choose for your discs, make sure that it meets these three criteria:

- It must be closed to dust and dirt.
- It must protect your discs from scratches.
- It must be expandable (or at least store enough discs to provide for some future additions)!

Cleaning Discs the Right Way

Although there are expensive DVD disc cleaning machines, I don't recommend them, and they're really not necessary—if you follow the Golden Rule, hold your discs correctly, and store them in their cases, they simply don't need cleaning!

However, accidents do happen. If a disc picks up fingerprints or dust, use a lint-free photographer's lens cloth—if necessary, you can also apply a spray or two of disc cleaning fluid if the disc has picked up a stain.

It's very important to wipe the surface of your discs properly: *Never wipe a compact disc in a circular motion!* (Figure 1.10) Wiping in a circle is more likely to scratch a larger portion of the disc's surface and result in read errors; always move from the center spindle hole straight toward the outside of the disc (Figure 1.11), applying only fingertip pressure to avoid scratches.

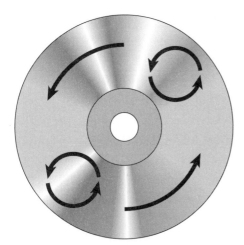

FIGURE 1.10 Never wipe a disc with a circular motion!

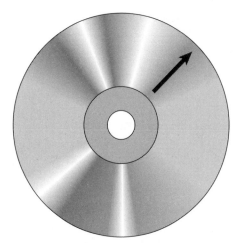

FIGURE 1.11   To avoid scratches, wipe discs from the center toward the outside of the disc.

A Word about Copy-Protected Discs

DVD-Video movies are practically impossible to copy: Some discs are protected by the addition of unreadable areas on the disc during manufacture (because your DVD recorder can't read these areas, it also can't duplicate them). Other protection schemes involve encrypted key codes that your DVD player must recognize in order to play the disc. Sure, it's possible to copy the digital video stream from a DVD disc, but it takes hours of time, specialized software, and quite a bit of technical knowledge; the quality of the video is likely to be significantly degraded, too. (I don't think I need to underline the fact that it's also illegal.) There are programs that advertise that they can "copy" DVD movies, but the result is far lower in quality than the original video (and you don't get any of the extras, such as interviews, subtitling, and alternate soundtracks).

But I'll be honest—there's a much more mundane, low-tech reason why copying DVD movies is a waste of time. Currently, a blank DVD-R disc typically costs around $2, but the hassle involved in producing a low-quality dub of a DVD movie can tie up your computer for a day or two (yes, it actually takes that long). If you can buy a legal copy of a movie for $15, is it really worth it?

To sum it up: A DVD recorder is perfect for recording your own home video, but don't expect to use it to copy commercial movies.

DVD Formats on Parade

As I close this chapter, I'd like to introduce you to the three DVD *formats* you'll be using with your new recorder. A format is a standardized "road map" used by hardware manufacturers around the world that details what data goes where on a recorded disc—without the proper format, your disc would be unreadable in any other device, so you couldn't use it in your DVD player or your computer's DVD-ROM drive. I'll try to keep the technobabble to a bare minimum while we cover these three important formats!

DVD-V

A commercially manufactured DVD movie disc is often called a DVD-ROM (that's short for *DVD read-only memory*)—technically, however, a DVD-ROM can also be called a DVD-V disc, which is short for *DVD video*. The DVD-V format describes a DVD-ROM that holds broadcast-quality digital video in a special compressed format called MPEG-2. As I mentioned earlier, DVD-V discs can also include high-tech features such as Dolby audio, surround sound, subtitles, and different aspect ratios—and you can even run programs from a DVD-V disc if it's being played in a computer DVD-ROM drive.

If you're using a DVD recorder, it is possible to record a DVD-V disc that delivers the quality of a commercially made DVD-ROM—in fact, I'll be showing you how to do just that later in the book, using the HP DVD-Writer and a selection of home movies I've taken with my DV (*digital video*) camcorder.

DVD-A

Here's a riddle for you: What looks like an audio CD but features up to 4 hours of stereo music, interactive animated menus (much like the current crop of DVD-V discs), and even the ability to store video clips? The answer is DVD-A, which stands for *DVD audio*! You also get the mind-boggling addition of surround sound, which provides unparalleled realism and depth to your music. DVD-A discs will not be compatible with your current audio CD player, but your collection of traditional audio CDs should work fine in a DVD-A player.

The future of DVD-A is somewhat cloudy, because it may take a number of years to match the popularity of the familiar audio compact disc, and it's now receiving stiff competition from the new Super Audio CD (or SACD) format that's currently in development.

UDF

Our last—but by no means least—recording format is called UDF (short for *Universal Disc Format*). You're also likely to hear UDF called by another common name, *packet-writing*, which dates back to the days when it first appeared in CD recording software. UDF recording allows you to write files to a DVD-R/DVD-RW, DVD+R/DVD+RW, or DVD-RAM just as you would write files to your hard drive, using Windows Explorer, the Mac OS Finder, or any of your favorite applications. If you can write the file to your hard drive or a floppy, you can write it directly to DVD. Files can be added (or deleted) one at a time or in groups, without having to record the entire disc at one sitting. In effect, UDF turns your DVD recorder into a gigantic, superfast, removable disc that can store at least 4.7 GB of data. Figure 1.12 illustrates the formatting utility you'll use with the popular UDF recording program DirectCD, which ships with Easy CD Creator; HP DVD-Writer owners can use the HP DLA program, which I'll be covering later, in Chapter 5 ("Drag-and-Drop Recording with HP DLA").

Later in the book, we'll encounter a number of other different formats that apply to computer CD-ROMs and audio CDs; your drive can record these formats, as well, and we'll use them for several projects.

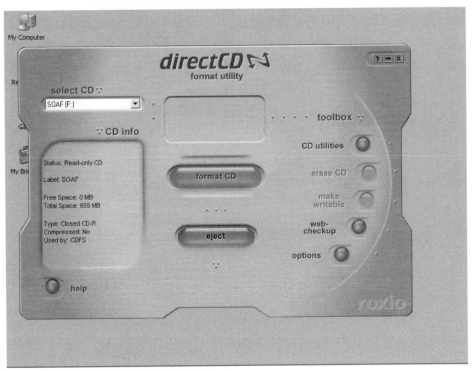

FIGURE 1.12   Preparing a disc for UDF recording using the DirectCD formatting utility.

Summary

This chapter introduced you to the basics of DVD recording, beginning with the binary language of optical storage. I discussed how data is burned and retrieved from a recordable DVD disc. You learned about the different types of DVD media, including their strong and weak points. I covered the proper methods of handling and storing your discs, and touched on the topic of copy-protected DVD discs. Finally, I discussed the three recording formats you'll use when creating DVD discs.

Installing Your DVD Recorder

In This Chapter

Most internal CD and DVD recorders available for today's computers use an EIDE connection—for technical types, that stands for Enhanced Integrated Drive Electronics, a type of connection that's supported by all of today's PCs. Luckily, however, normal human beings need only remember the acronym (and only when shopping, just to make sure that you've bought a drive with the right kind of connection).

Although EIDE hardware is very easy to install, there are still jumpers to configure, cables to connect, and a drive that needs to find its way into your computer's case—and that's what this chapter is all about. I'll show you how to install both your drive and the bundled software that came with it…and don't worry, you don't have to be a computer technician to install a DVD recorder. As a matter of fact, if you can handle a screwdriver, you already have all the tools you need!

By reading the instructions that came with your recorder and by following the step-by-step procedures in this chapter, *anyone* can install their own drive (and brag about it afterward).

If you've decided to add an external drive to your system using FireWire or USB, things are much easier: You won't have to remove your computer's case, move jumpers, or worry about "which end of the cable is up." However, there are still one or two possible pitfalls you may encounter, so I'll also cover the connection of an external recorder.

Before We Begin

In this chapter, I'm going to demonstrate how to install both the Hewlett-Packard DVD-Writer DVD200i—an internal EIDE device—and an external FireWire or USB drive such as the HP DVD200e. Although it's harder to install an internal drive than an external recorder, the step-by-step procedure in this chapter should be all you need. If you have another brand of EIDE DVD recorder, *you don't have to skip this material!* EIDE devices are very standardized (they have to be to be so compatible); therefore, the steps you'll read here will also work with virtually every combination of EIDE DVD recorder and a PC.

There is the possibility that your drive may require special software or use a different jumper configuration than the DVD200i recorder. Because of the chance that you'll encounter something different, I strongly recommend that you read the installation instructions for your recorder! Once you're familiar with the installation instructions for your particular drive model, you can follow the steps I provide in this chapter—and if the installation instructions for your drive differ significantly from mine, you can follow them instead.

What Do I Need?

Let's consider the basic requirements for installing your new DVD recorder; if your PC doesn't meet these minimum standards, you'll have to upgrade your system before you can use your drive. An EIDE recorder requires:

- **An EIDE cable with an open connector.** Most PCs built in the last five years or so have four available EIDE device connectors: a *primary* master and slave on one cable, and a *secondary* master and slave on another cable. (If that sounds like Greek to you, don't worry—you'll learn more about master and slave settings later in this chapter). Your new drive requires one connector. If your PC has only one hard drive and one CD-ROM drive on the primary cable, you may have to buy another EIDE cable for your secondary connector; many PC manufacturers don't provide a cable if the secondary connector isn't being used.

- **An empty drive bay.** Your case must have one 5.25-inch drive bay available that's open to the front of your case; it's the same type of drive bay your CD-ROM drive uses. Most PC cases have at least two of these bays.

- **An unused power connector.** Your new recorder needs an available connector to your computer's power supply, such as the one illustrated in Figure 2.1. If your computer no longer has a spare power connector, you can buy a power splitter (also called a *Y connector*) at your local computer store; these special plugs convert one power cable into two.

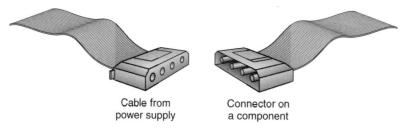

Cable from
power supply

Connector on
a component

FIGURE 2.1   Your PC's power supply must have an available power cable to
install your drive.

If you're installing an external FireWire or USB drive, you'll need
the following:

- **An open connector.** At the time of this writing, there's only
one standard for a FireWire (also called IEEE-1394) connection,
so you can be reasonably sure that you've got the right hardware. FireWire is used for everything from high-speed external
hard drives and recorders to DV cameras and scanners. Things
aren't quite as simple on the USB side of the fence, however,
because there are now two versions of the Universal Serial Bus:
the original standard, which has been around since the late
1990s, and the much faster version 2.0 standard. Luckily, version
2.0 cables and devices are backward-compatible with original
USB hardware, so everything will still work—your connection
won't be anywhere near as fast, though, and you probably won't
be able to record at the full speed possible with your drive.
Therefore, if you have a recorder that supports USB 2.0 (such
as the HP DVD200e), make sure that the USB ports on your
computer conform to the new 2.0 standard so that you can take
advantage of the faster data transfer rate. (Many manufacturers
are already selling USB 2.0 adapters for PCs, so you can add
these ports to your computer for about $150.)
- **An open AC wall socket.** Unlike an internal drive (which
draws its power directly from your PC's power supply), an
external recorder has its own power cord.

tip Like any other piece of expensive external hardware—your printer, for example, or your scanner—you should connect your external recorder to a surge protector. If you have the cash to spare, a UPS (or Universal Power Supply) is even better: Not only will you protect your system from brownouts and surges but, in case of a power outage, you can continue to run your computer for a number of minutes. If you're recording a disc when the lights go out, those extra minutes can mean the difference between a successful recording and a disc that's headed for the trash can!

- **An appropriate location.** Find the right spot for your external recorder! Keep it several inches away from your computer's case and monitor, which ensures proper ventilation, and make sure the drive tray has the room to extend fully. Also, make sure that it's safe from vibration; for example, don't place that sensitive drive on top of your PC's subwoofer!

Next, let's turn our attention to the requirements you'll need for a typical suite of DVD recording software. Your PC should have a minimum of the following hardware:

- **128 MB of RAM.** Today's recording software needs elbow room to work. If your computer has less than 128 MB of RAM, take advantage of today's low memory prices and upgrade your PC; more memory will allow all versions of Windows and your application programs to run faster, too.
- **Pentium III or faster CPU.** For error-free video recording to a DVD drive, I recommend a minimum of an 800-MHz Pentium III computer; anything slower, and your computer will likely not be able to record successfully. (For simple data recording, you can get by with a minimum of a 450-MHz Pentium II computer.)
- **Windows 98, Me, XP, or Windows 2000.** HP supports Windows 98 Second Edition, Me, 2000, and XP (Windows NT and the advanced server versions of 2000 and XP are not supported). If you're using Windows 95, it's definitely time to upgrade!

- **5 GB of free hard drive space.** Wow! Those DVD+RW discs can really hold a huge amount, but you typically have to store all that data on your hard drive first. If you want to record your own DVD video and your hard drive doesn't have 10 GB available, you'll have to either delete existing software to make room or buy a new drive (either to add to your system or to upgrade your current hard drive). For regular data recording to a CD-R or CD-RW, it's good to have at least 1 GB free.

- **Video capture hardware.** To create DVD-V discs using your own video clips, you'll need either a FireWire port on your PC or a video capture card.

Be Prepared

It works for Boy Scouts, and you'll find that good preparation can also help keep your installation short and smooth. I recommend the following preparations before you pick up your screwdriver:

- **Set up a stable surface, such as a tabletop or workbench.** If you need to protect the top of your work surface, such as the kitchen table, spread newspaper on top. Do not use cloth, which can result in static electricity!

- **Keep a good light source handy.** You can't work on something that you can barely see; I use an old-fashioned gooseneck desk lamp on my workbench.

- **Use a parts bowl.** What's a parts bowl? Nothing exotic—just a plastic bowl or wide-mouth container that can hold all of the screws and parts that you remove (either temporarily or permanently). Once the installation has been finished, anything left over in the bowl can be saved for future installations and repairs; you'll be surprised at just how good it feels to have the right spare part!

- **Reserve plenty of time.** No one wants to be rushed while installing computer hardware: You're more likely to overlook a step or leave a cable unattached. If you've never installed a drive before, allow yourself at least three or four hours of uninterrupted time.

> tip **Do you have a friend who's knowledgeable about computer hardware? If you'd like a "safety net" while you install your first EIDE drive, it's a good idea to enlist that person to help.**

Time to Determine Your Settings

As I mentioned earlier, most PCs can handle up to four EIDE drives, with two drives on each of two cables. How does your PC keep track of which devices are connected? By the use of jumper settings. A *jumper* is a tiny wire-and-plastic electrical crossover that connects pins on your recorder's circuit board. You can move a jumper to different positions to connect different sets of pins, which in turn changes the drive's EIDE configuration. (Tweezers come in handy if the jumper is particularly small.)

Determining Jumper Settings for an HP DVD-Writer

Hewlett-Packard includes a software utility with the DVD-Writer DVD200i drive that can recommend which jumper settings you should use. To use this program, you should install the software that came with the DVD-Writer now, before you even open your computer's case!

✔ **Follow these steps:**

1. Load Software Disc 1 into your PC's CD-ROM drive, and you'll see the screen shown in Figure 2.2.

2. Click Start Installation to begin, and follow the on-screen instructions to install the various programs that you received with your DVD-Writer.

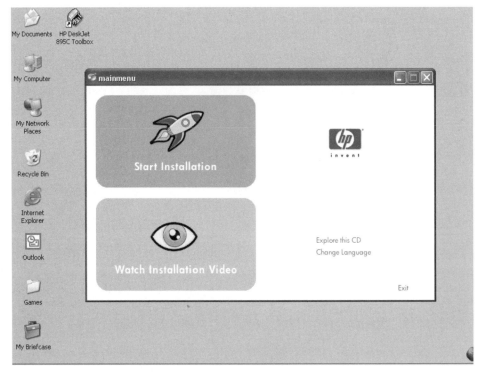

FIGURE 2.2   The Welcome screen from the HP DVD-Writer software installation
program.

tip **During the software installation, it's a good idea to accept
all of the defaults (especially which programs you'll install and
where they'll be stored)—this ensures that you'll have all the software
you'll need for many of the projects I'll be showing you later. Also, you
should install DirectX8.1 when prompted, because you'll need it for
advanced video work.**

3. Once the software has been installed, you'll see the screen shown
    in Figure 2.3—the software has recognized that you haven't yet
    installed the drive itself. Click Next to continue.

4. The installation program displays the screen illustrated in Figure
    2.4, where you'll decide whether to keep your PC's existing CD-
    ROM or replace it with the DVD-Writer. (Personally, I recommend
    that you keep your existing CD-ROM drive if you have an open bay

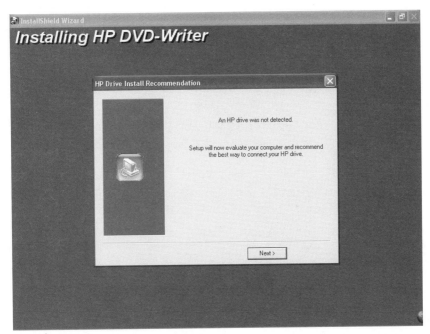

FIGURE 2.3 Beginning the Install Recommendation portion of the installation.

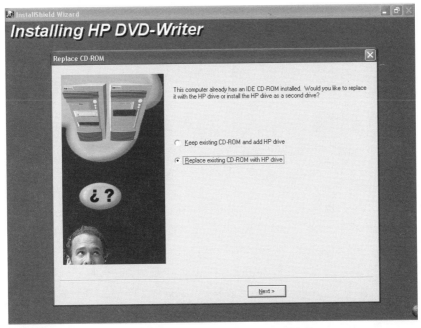

FIGURE 2.4 To add or switch—that is the question!

available. Of course, if you don't have a spare open bay, you'll have to replace your CD-ROM drive.) Click the desired option and click Next to continue.

5. The program displays a customized screen like the one in Figure 2.5 that shows you the correct jumper settings, cable, and EIDE connector you should use. Write down this information, or click Print to send it to your system printer. Click Next.

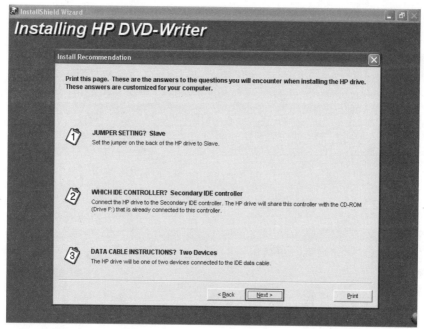

FIGURE 2.5   The installation program displays the settings you'll need for your hardware.

6. The final installation screen you see in Figure 2.6 offers several options. Click Review the drive Install recommendation to return to the Recommendation screen shown in Step 4. Click Watch the tutorial install Video to see a 10-minute animated video overview of the hardware installation process—I highly recommend watching this video, it's very well done, and it's a good preparation for the steps to come. To begin the hardware installation process immedi-

ately, click Shutdown Now; to install the hardware later, click Shutdown later (remember, however, that you *must* install the drive before you can restart your PC again, because programs that check for the drive have already been installed).

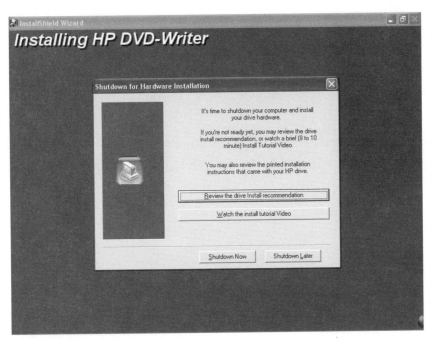

FIGURE 2.6 Finishing the software installation.

If you decide to shut down later, you'll see two new icons on your Windows desktop, as shown in Figure 2.7: PowerDVD and HP DVD-Writer!

FIGURE 2.7   Your desktop sports new icons.

Determining Jumper Settings for Other DVD Recorders

If you're using another brand of drive, you'll probably have to deter-mine the jumper positions you should use for your existing EIDE hard drive, your existing CD-ROM or DVD-ROM drive (if you have one), and your new recorder. Table 2.1 should help you decide which set-tings are correct; the different "master" and "slave" positions are printed on most EIDE drives, and they can also be found in your hardware documentation.

tip **If all else fails and you can't locate the jumper settings for an older hard drive or CD-ROM drive, visit the manufacturer's Web site and check its technical support area, or call its voice technical support number for help.**

TABLE 2.1   EIDE Master/Slave Jumper Settings

One Hard Drive	One Hard Drive and One CD-ROM on the Same Cable	One Hard Drive and One CD-ROM on Different Cables
Set the hard drive as "multiple drive, master unit" and set your recorder as "multiple drive, slave unit."	Set your recorder as "single drive, master unit" and connect it to the secondary EIDE cable. No changes are required to the existing drives on the primary EIDE cable.	Set your recorder as "multiple drive, slave unit" and connect it to the EIDE cable that the hard drive is using. The hard drive must be set to "multiple drive, master unit." No changes are required to the CD-ROM drive on the secondary EIDE cable.

Installing an Internal Drive

Ready to install your new DVD recorder? Is your work surface prepared, well lit, and equipped with a parts bowl? In this section, I'll take you step by step through the entire process.

caution   Static electricity can damage any exposed electronic components you touch, including your PC's motherboard and adapter cards! Therefore, always take a moment to discharge any static electricity before touching the interior of your PC or any hardware—touch a nonpainted metal surface (such as the back of your computer or your computer's case) often before and during the installation.

✔ **Follow these steps to install an internal drive:**

1. Shut down your computer and unplug it from the AC wall outlet.

2. Unplug all of the various cables from the back of your PC (if necessary, take a moment to label them or draw a diagram of the cable connections before you unplug them).

3. Remove the screws that secure the cover to your computer and put them in your parts bowl. Take off the cover and set it aside.

4. Set the jumpers on the back of your recorder according to the recommendations made by the HP software or your drive's documentation. Figure 2.8 shows the three possible settings for the HP DVD-Writer drive. (Remember, you may have to change the settings for other EIDE devices that are already installed in your PC—if so, make those changes now, as well.)

5. Select an open drive bay for the recorder that offers access to the front of the case. If the bay is covered by a plastic insert, remove it by pushing on it from inside the case, as shown in Figure 2.9. Some of these inserts are snapped into place, so you may have to bend the insert to remove it; if so, be careful not to gouge or scratch your case in the process.

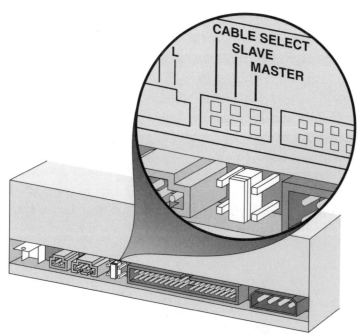

FIGURE 2.8   Setting the master/slave jumper on an HP DVD-Writer drive.
Image courtesy of Hewlett-Packard.

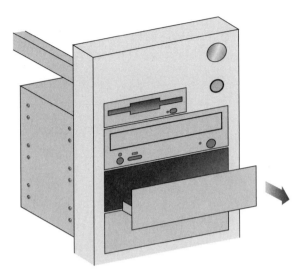

FIGURE 2.9   Removing the plastic insert covering a drive bay. *Image courtesy of Hewlett-Packard.*

6. Once the bay is open, slide the drive into it from the front of the case, as shown in Figure 2.10. The end with the connectors should go in first, and the drive should be facing upright (check the printing or lettering on the front of the recorder to make sure that the orientation is correct).

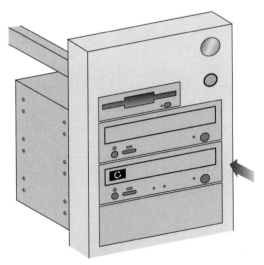

FIGURE 2.10   Introducing the drive to its new home. *Image courtesy of Hewlett-Packard.*

7. Slide the recorder back and forth in the drive bay until the screw holes in the side of the bay line up with the screw holes on the drive. Attach the recorder to the bay with the screws supplied with the drive, as shown in Figure 2.11. Although you should try to add all four screws, it may be harder to reach some of them (with at least two screws, the recorder will be fine).

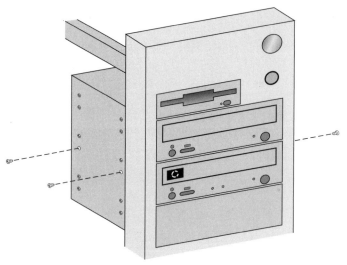

FIGURE 2.11   *Securing the drive. Image courtesy of Hewlett-Packard.*

8. Connect a power cable from your PC's power supply to the recorder's power connector, as shown in Figure 2.12. Don't be worried about plugging the power cable in the wrong way—the connector fits only one way, so there's no chance of making a mistake! (I'd like to thank the engineer who designed these.)

9. Now it's time to attach the EIDE data cable coming from the computer's EIDE connector to the data connector on the back of the recorder, as shown in Figure 2.13. If your new drive will use a cable by itself, you may have to connect the cable to your computer's motherboard, as well—check your motherboard manual to determine where the secondary IDE connector is located. If your recorder will share a cable with an existing EIDE device, simply fol-

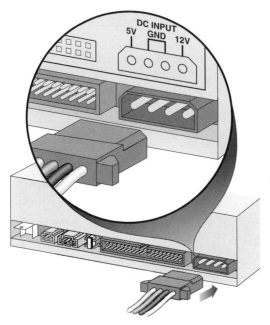

FIGURE 2.12 Providing power to your new recorder. *Image courtesy of Hewlett-Packard.*

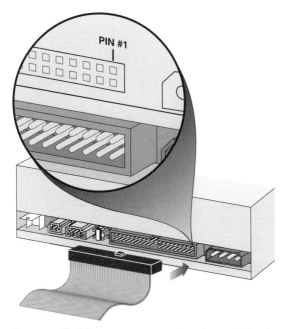

FIGURE 2.13 Connecting the data cable. *Image courtesy of Hewlett-Packard.*

low the cable from the back of that device until you find the second connector. Again, this cable should fit only one way; a small notch on the cable connector must line up correctly with a matching cut-out on the drive's connector, and Pin 1 normally lines up with the side with the power connector. Once you're sure that the alignment is correct, press the connectors firmly together.

10. If you plan to play audio CDs using your DVD-Writer, you must connect the cable from your sound card to the port on the back of your drive, as shown in Figure 2.14. If you're going to keep your existing CD-ROM drive, this cable is probably already attached, and you can either move it to your DVD recorder or leave it connected to the CD-ROM. (If you choose the latter, you should continue to listen to audio CDs with your CD-ROM drive.) PCs manufactured by HP usually ship with sound cards that have two audio cables; in this case, use both.

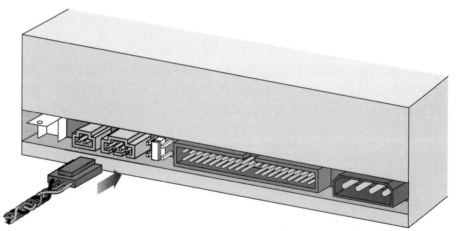

FIGURE 2.14 Connecting the audio cable. *Image courtesy of Hewlett-Packard.*

11. Check all connections to each device (even the cables you didn't touch)—it's very easy to inadvertently unplug a cable while working inside your PC.

12. Replace your PC's cover and fasten it with the original screws from your parts bowl.

13. Plug all of the cables from your other devices back into your PC.

14. Plug in the power cable and turn on your computer.

Remember, if you're installing a DVD recorder other than an HP DVD-Writer, you may have additional software that you have to install.

Installing an External Drive

If you've bought a USB or FireWire external drive, there's no need to disassemble anything...in fact, you don't even have to turn off your PC!

✔ **Follow these steps to connect the drive to your computer:**

1. Connect the drive's power cord to the AC wall socket.

2. Connect the USB or FireWire cable to the drive and turn it on.

3. Plug the cable from your drive into the corresponding USB or FireWire port on your computer.

4. If you're connecting the drive for the first time, load the driver disc when prompted (or install the software that came with your drive).

Once you've successfully connected the drive for the first time and loaded any required software, you can put the manufacturer's disc away—it's necessary only when loading software the first time.

Troubleshooting EIDE Installation Problems

Although EIDE drives are the industry standard, there's still the possibility that your new recorder will sit motionless like a silicon brick when you try to use it—if so, it's time to exercise your troubleshooting skills to determine the problem and fix it. Never fear, I won't abandon you in your time of need! I'll cover the most common installation glitches you may encounter and their solutions.

My new recorder doesn't eject or show a power light.

- **Power cable loose or unplugged.** If the drive's tray doesn't eject when you press the Eject button, check the power connector—even if the jumper settings aren't correct, your drive should still be able to eject the tray, as long as it's receiving power.

Windows no longer boots—my hard drive is no longer recognized.

- **Cables improperly connected.** This can happen if you've accidentally moved your PC's hard drive from the primary to the secondary controller cable—the hard drive must be the master device on the primary EIDE controller cable.

I get power, but Windows doesn't recognize my new drive.

- **Master/slave jumpers are incorrectly set.** This is a classic symptom of an improper jumper setting. If your new DVD recorder is sharing a cable with another EIDE device, make sure that you've correctly set the master/slave jumpers for *both* drives.
- **Data cable loose or unplugged.** Check the ribbon cable connection to both the drive and the motherboard, and make sure that the cable is firmly seated on both ends. (If the drive is sharing a cable with another EIDE device and that device is working fine, the culprit will be the connection to the DVD recorder.)
- **Required software drivers haven't been loaded.** Check to make sure that you've run the complete software installation process for your new drive.

I get a weird message about something called a BIOS error.

- **Incorrect BIOS configuration.** PCs made within the last three or four years will automatically detect changes in your

EIDE setup and add (or remove) devices from your BIOS configuration—if your PC doesn't do this by itself, however, you can make the required changes manually. Check your PC's documentation for the key that you need to press during the boot sequence to edit your system BIOS—when you've displayed the BIOS, use the update function to add the recorder to your active EIDE device list.

Troubleshooting USB and FireWire Installation Problems

You've already seen that USB and FireWire connections are about as trouble-free as you can get—but, like every other piece of hardware on the planet, it's possible that things may not work as advertised. In this last section, I'll discuss common problems with external drives and the proper way to handle them.

My new recorder doesn't eject or show a power light.

- **Power cable loose or unplugged.** Just like an internal recorder, your new external drive needs power! Check the AC cable and make sure that it's snugly connected to the drive.

tip Some USB 2.0 drives may not use a separate power supply—instead, they may draw power directly from the port itself. Unfortunately, not all USB ports can provide that much juice (some are provided only for connecting a keyboard and a mouse); this is the likely problem if you find that your drive works fine when connected to one computer but doesn't seem to be getting any power at all when connected to another system.

I get a message about a missing driver.

- **USB/FireWire software drivers corrupted, incompatible, or not installed.** The culprit is the manufacturer's driver program: It may have been damaged, or you may not have installed it on this computer. Alternately, your recorder's driver may not work with your operating system (a classic problem that often occurs when you've just completed an upgrade to your operating system). Try reinstalling the manufacturer's software—if the driver doesn't work with your operating system, it's time to visit the drive manufacturer's Web site to hunt for an updated driver.

I get power, but Windows doesn't recognize my new drive.

- **Cabling problems.** This problem is usually traced to a faulty cable or a faulty hub (a separate piece of hardware that allows you to connect multiple USB or FireWire devices to a single port). Try connecting the drive with a different cable.
- **No power when daisy-chaining.** If you've connected your recorder to a USB or FireWire device that has a "passthru" or "daisy-chain" port, make sure that *both* external devices are switched on! Most USB and FireWire hardware won't pass a signal to further devices if they're turned off.

Summary

In this chapter, you installed your new DVD recorder and its software. I discussed the requirements for adding both an internal and an external DVD recorder to your PC, as well as the preparations you should take before beginning the installation. I then provided step-by-step instructions, using the HP DVD-Writer as an example. Finally, I discussed the most common problems you might encounter during the installation of an EIDE, USB, or FireWire device, and I provided possible solutions for each problem.

3

Preparing Your Computer and Material

In This Chapter

✓ Reducing hard drive clutter

✓ Defragmenting your hard drive

✓ Preventing interruptions from scheduled programs

✓ Adding memory to improve performance

✓ Converting files to archival formats

✓ Organizing your files for easy retrieval

Many folks ask me, "Can't I just record whatever's on my hard drive?" The answer, of course, is yes—once your DVD recorder and its software have been successfully installed, you're "technically" ready to burn discs. However, I don't recommend that you rush right into recording without taking a few simple steps first—without this preparation, you may:

- *Run out of space to hold the files you want to record (or your recording software may run out of temporary file space in the middle of a burn, which is even worse).*

- *Degrade the performance of both your recorder and your entire system.*
- *Record documents and files that you may not be able to read five years in the future.*
- *Record files haphazardly, making it harder to find the files you're looking for when you need them the most.*

I n this chapter, I'll show you the preparations you should take before you record your first disc—these steps will reduce the possibility of recording errors and speed up the burning process. You'll also find tips and tricks you can use to organize and standardize your files…it's a good bet that you'll benefit from these preparations the next time you load one of your discs!

Freeing Up Space on Your Hard Drive

Today's hard drives provide dozens of gigabytes of storage, but they aren't limitless. Today's 3D games often require 700 MB of space, and a lengthy digital video clip can easily gobble a gigabyte, so it's actually more important than ever to make sure you've got enough open territory on your hard drive! Besides the space required to hold your material, you also need free space because:

- Your recording software will need additional space for temporary files; these files are deleted once the recording has finished.
- Your computer may need space to convert files from one format to another (such as WAV audio files to MP3 audio) or compress digital video before it can be recorded.
- Windows will need additional space for *virtual memory* (hard drive space Windows uses as "temporary RAM" when your PC runs out of memory). Figure 3.1 illustrates how virtual memory works.

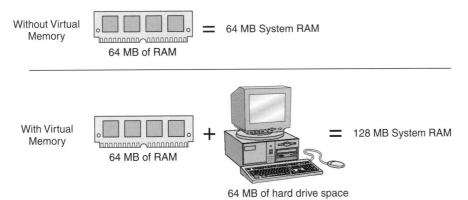

Without Virtual Memory   = 64 MB System RAM

64 MB of RAM

With Virtual Memory   +   = 128 MB System RAM

64 MB of RAM

64 MB of hard drive space

FIGURE 3.1   With virtual memory, a PC can run programs that use more memory.

Naturally, you won't want to simply drag a gigabyte's worth of files straight to the Recycle Bin! In most cases, you can safely delete these types of files:

- **Files stored in your Recycle Bin.** Those files stored in your Recycle Bin are actually still taking up space on your hard drive, and because you've already deleted them, it's safe to release that space. Right-click the Recycle Bin icon; Figure 3.2 illustrates the pop-up menu that appears. Select Empty Recycle Bin to delete the files.

- **Windows temporary files.** Windows 98, ME, and XP create all sorts of temporary files that can clutter up your hard drive. To clean up these derelicts, use the Windows Disk Cleanup Wizard. Under Windows XP, for instance, you would click Start | All Programs | Accessories | System Tools | Disk Cleanup, which runs the Wizard you see in Figure 3.3.

tip If you're familiar with the XP Disk Cleanup Wizard, you may have noticed that it offers you the opportunity to compress older, seldom-used files to regain space. This is a good idea—on a larger drive of 40 GB or more, compression is likely to save you at least 1 GB—and it won't affect your recording performance.

FIGURE 3.2   Need more space? Empty your Recycle Bin.

- **Unnecessary programs, game demos, shareware, and sample files.** Whether the programs are applications that you no longer use, forgotten game demos, or shareware you decided not to buy, they're prime targets for removal. Also, if an application has example files or sample documents that you know you won't need, you may be able to trim them—check the program's documentation or README file for details.

caution   Never simply drag a program's folder to the Recycle Bin—instead, click Start | Control Panel | Add and Remove Programs (Figure 3.4)! If you remove a program without allowing Windows to uninstall it, you may cause your PC to lock up or cause other programs to stop running. Older Windows programs may have their own uninstall programs in the Start menu, as well.

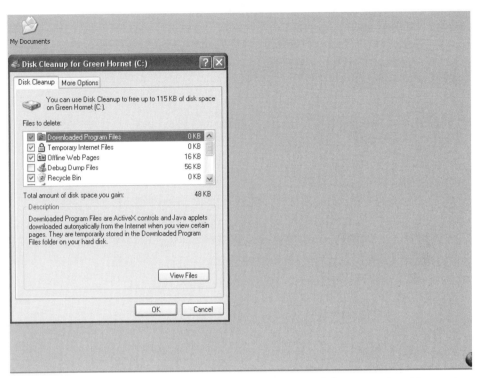

FIGURE 3.3   The Disk Cleanup Wizard may be buried in menus, but it's worth the trip.

- **Browser cache files.** If you spend hours browsing the Web, Internet Explorer may be eating up dozens of megabytes with its temporary files! To get rid of these files, run Internet Explorer, click Tools, and choose the Internet Options menu item. Click the General tab on the Internet Options dialog to display the settings you see in Figure 3.5. Click Delete Files, and click OK on the confirmation dialog.

If you can't find enough space on your system to use your new DVD recorder, it's time either to swap your existing drive with a larger one or to add a second drive to your system. Luckily, an EIDE hard drive is installed just like your DVD recorder, so you can follow the procedure in Chapter 2 in the section titled "Installing an Internal Drive" (and you won't need a device bay that's open to the front of the computer, because a hard drive has no tray to eject).

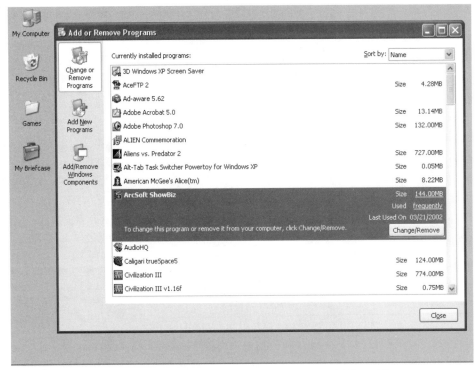

FIGURE 3.4   Deleting Windows applications using the Add and Remove Programs dialog.

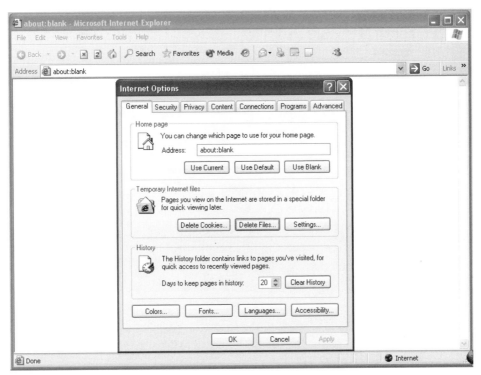

FIGURE 3.5 Deleting those pesky Web browser temporary files.

Defragmenting Your Hard Drive

Defragmenting is an easy step that you can take that will boost performance on any PC—and it can help eliminate DVD recording errors on older computers!

What is fragmentation, and why is it a problem? To answer those questions, you'll have to return to the days when you first began using your computer's hard drive. At that time, files were written contiguously (all in the same area and unbroken from beginning to end). As your drive stored more data, however, Windows started to save new files in the space freed when you deleted other files: Larger files can no longer be saved contiguously. Instead, a larger file is broken up into smaller pieces that fit into those "holes" left by deleted files, and it is saved in pieces (or segments) on your drive. The segment locations are saved, so your drive knows where all the parts of a particular file are stored. When a program wants to read that file, Windows and your drive's controller work together automatically to combine all the smaller segments of the file back into their original order; then Windows sends the restored file to the program.

It's a pretty neat process and invisible to both you and the programs you run—but the process isn't perfect. As you can see in Figure 3.6, it takes more time for Windows to "reassemble" the file, and the delay grows if there is a large number of segments that are spread out all over your drive. This is the ongoing process of fragmentation, and it gets worse over time—more files become more and more fragmented, and they take longer and longer to reassemble, which translates into slower disk performance. Disk performance is critical when you're recording a CD or DVD on a slower computer (especially when you're using Direct-to-Disc recording, as I will demonstrate in Chapter 7, "Direct-to-DVD Recording").

To optimize your drive, you need a defragmenter: a utility program that reads in the files on your drive and rewrites them in contiguous form, one after another. Figure 3.7 shows that same file after a defragmenter has done its work—our file is now contiguous, so it takes far less time for Windows to read it and send it to your recorder. It also means there's less chance that the next file you save to disk will become fragmented.

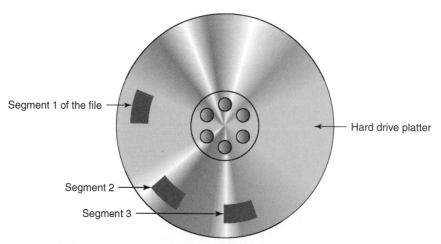

FIGURE 3.6   A fragmented file actually resides in segments on your hard drive.

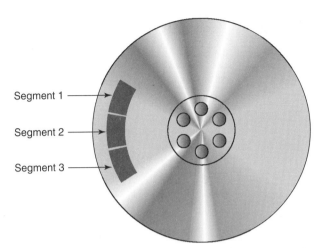

Segment 1

Segment 2

Segment 3

FIGURE 3.7   After running a defragmenter, the file is contiguous and loads much faster.

A defragmenter has been included with Windows 98, ME, and XP, as well as with Windows 2000—unfortunately, many folks don't know it's there and have never run it! I recommend running this program at least once a month.

✔ **To run Defragmenter under Windows XP, follow these steps:**

1. Click Start | All Programs | Accessories | System Tools | Disk Defragmenter. The Disk Defragmenter screen shown in Figure 3.8 appears.

2. Click the hard drive you want to scan. (If you have more than one drive, choose the drive that holds the files you'll be recording; however, it's always a good idea to defragment all of your hard drives.)

3. Click Defragment to begin defragmenting your drive. If you like, watch as Defragmenter does its thing—once the program has completed, you'll see by the graph that virtually all of the files on your drive are now contiguous.

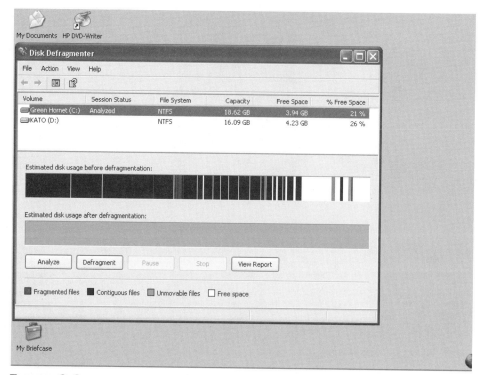

FIGURE 3.8   Windows XP includes a defragmenter—thanks, Microsoft!

Disabling Scheduled Programs

Not every Windows program is "well behaved." (Programmers would love you to think so, but it's not true.) Some of your favorite applications may be downright obnoxious; for example, a misbehaving program may prompt you for input without allowing other programs running simultaneously to continue in the background (effectively halting your entire system). Other programs may simply lock up from time to time. These types of interruptions during a recording session likely spell doom for that disc—if you're recording a CD-R, another ruined coaster hits the bottom of your trash can. (CD-RW and DVD+RW discs can simply be reformatted.)

To make sure your recordings aren't interrupted, you should disable any program that may kick in during a recording session, as well as other programs that may be running minimized or in the background. If a program isn't necessary to Windows or your recording software, disable or exit that program before you launch your recording software—and that includes programs running under Windows 2000 or XP Professional.

These programs can include:

- **MP3 or DVD players.** It's never a good idea to access any type of multimedia while you're recording. For example, playing a movie in Windows Media Player (shown in Figure 3.9) usually entails a great deal of hard drive and DVD-ROM access, and some MP3 players can cause your PC to freeze for a second or two at the beginning or ending of a song.

FIGURE 3.9   Avoid the popcorn—never watch a DVD while recording.

- **Disk and system monitoring software.** It's important to scan your hard drives for errors, but do it before you start recording, not *during*! Also, system monitoring programs, such as Norton System Doctor (Figure 3.10), are great for checking the overall performance of your PC, but they're also infamous for slowing down older computers and automatically running programs (neither of which is recommended during a burn).

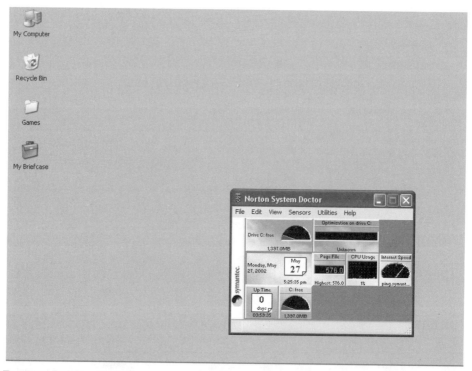

FIGURE 3.10   Disable monitoring programs, such as Norton System Doctor, while recording.

- **Screensavers.** Pretty to look at, but they can turn a fast Pentium III or Pentium 4 processor into a paperweight—especially multimedia screensavers that feature digital video or animation. Under Windows XP, right-click the desktop and choose Properties, then click the Screensaver tab (Figure 3.11) and set your screensaver to None.

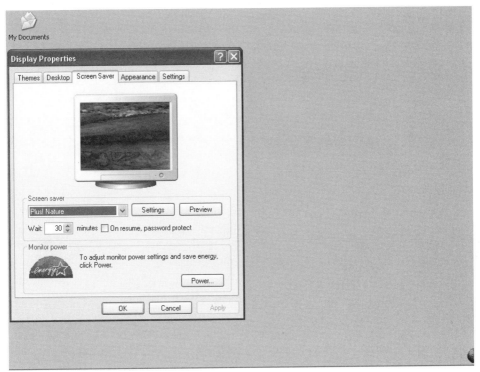

FIGURE 3.11   Turning off the screensaver in Windows XP.

- **Fax programs.** Fax programs can be performance-intensive, so be sure to disable your fax program's receive and scheduled send functions before you record.

You can usually disable or exit an unnecessary background task by right-clicking its taskbar icon and selecting Quit or Disable from the pop-up menu, as shown in Figure 3.12.

I also recommend pausing the Windows Task Scheduler (Figure 3.13) while you're recording—this prevents it from automatically launching programs without warning right in the middle of a burn. Right-click the Scheduler icon in the taskbar, and select Pause from the pop-up menu. (Don't forget to select Continue Task Scheduler after you've finished recording!)

FIGURE 3.12   Turning off an unnecessary background task.

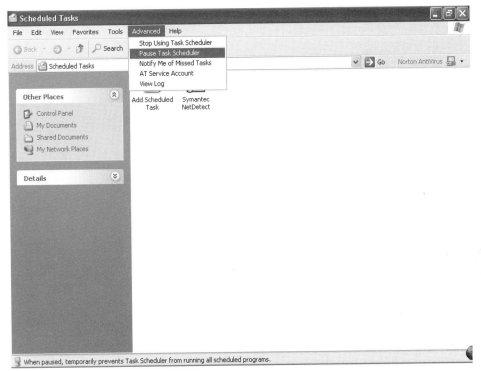

FIGURE 3.13 Pausing the Task Scheduler in Windows XP.

Adding System Memory

I'll be blunt here: At the time of this writing, memory is *dirt cheap*. If you're running Windows 98, I recommend at least 128 MB of RAM; if you're running Windows NT, ME, 2000, or XP, I wouldn't settle for less than a minimum of 256 MB of system RAM. If your computing budget allows more than these minimums, then by all means add more!

If you're unsure how much memory is available to Windows, right-click on the My Computer icon and select Properties from the pop-up menu. Figure 3.14 illustrates the System Properties dialog that appears in Windows XP; as you can see, this particular machine has 384 MB of RAM.

Why am I so enthralled with PC memory modules? Simple: No matter which version of Windows you're running, it will *always* per-

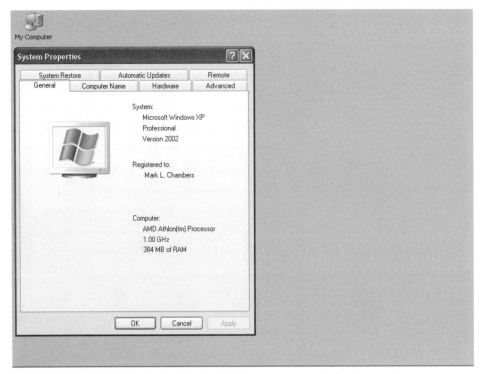

FIGURE 3.14 My machine has 384 MB of system memory.

form better overall with each extra megabyte of RAM you add. In fact, short of a processor upgrade, additional memory is the most effective upgrade you can perform when it comes to speeding up your system.

Also, adding more memory is very easy to do. If you like, you can bring your PC into your local computer store and have them perform the upgrade for you, but installing additional memory is much easier than installing your recorder; the entire process should take less than 10 minutes. Most of today's PCs use DIMM memory modules (there are a number of varieties, some of which are faster and more expensive than others).

✔ **To add a DIMM module to your PC, follow these steps:**

1. Shut down your computer.

2. Touch the metal chassis of your computer to dissipate any static electricity from your body, then unplug it from the AC wall outlet.

3. Unplug all of the cables from the back of your PC.

4. Remove the screws holding the cover and put them in your parts bowl. Take off the cover and set it aside.

5. Locate the DIMM memory slots; typically, they're located at one corner of the motherboard, or close to the CPU itself. Your motherboard manual should include a drawing that will help you find the memory slots.

6. Align the connector on the bottom of the memory module with the socket and push down lightly to seat the chip; make sure that the notches cut into the module connector match the guides in the memory sockets. (These guides ensure that you won't install your memory modules the wrong way!)

7. Make sure the two levers at each side of the socket move toward the center—these levers at the ends of the chip lock it into place.

When the DIMM is correctly installed, it should set vertically on the motherboard, with the two levers flush against the sides of the module.

tip **When buying memory for your PC, make sure that you're buying the right type (older computers may even require that you upgrade memory modules in pairs). Bring your PC's manual to the store, or check the manufacturer's specifications for the right type of memory to use. (If you're totally stumped as to what to buy, you can even remove the existing modules and bring them with you—make sure they're wrapped in a static-free plastic bag.) The manufacturer's Web site often carries memory upgrade information for discontinued PC models.**

Converting Files to Other Formats

First, a quick definition: If you've never heard of file *conversion*, it's the procedure you follow when you change a file from one format to another. Many computer applications understand only one or two file formats, whereas others can understand many—and that's the reason behind this section.

When you choose documents and files that will be saved on DVD and used for many years, it becomes important to save those files in the right format; this can help ensure that you'll still be able to watch your digital video, listen to your music, or open your Letter to the Editor in 10 years as easily as the day after tomorrow. Formats do "die out" over the years—this makes perfect sense, because everyone tends to standardize on one or two very successful and popular formats. After that decade has passed, you definitely want your music saved in MP3 format and not SND format. (Never heard of SND format? Gotcha! It's an older, once-popular Macintosh sound recording format that has all but disappeared, and it's the perfect proof of my point.)

There's another reason why you might convert files before recording them on a disc: If you're creating a cross-platform disc that will be read by Macintosh owners (or Linux technowizards), you'll save those people quite a bit of trouble if you choose a format that's just as popular in their computing world as in the Windows world. Don't leave an image bound for a Macintosh in Windows Bitmap format, for example—instead, do the job yourself and convert that file to JPEG or TIFF (both of which are popular in the Mac world and well recognized by Macintosh applications).

Here are the best choices for formats for each of the major types of files you're likely to record, along with the best programs for converting file formats under Windows:

- **Images and photographs.** Use either JPEG or TIFF—JPEG is the most widely recognized image format out there (and it's the format of choice for Web pages), and TIFF is almost as well known. TIFF images provide better quality than JPEG, but they're much, much larger. Use Paint Shop Pro from Jasc (Figure 3.15) for conversions; it's inexpensive shareware, it's fast, and it

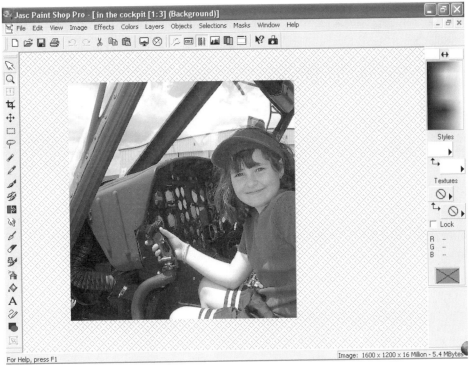

FIGURE 3.15 Converting images using Paint Shop Pro.

recognizes a huge number of formats! Jasc's Web site is located at www.jasc.com. Adobe Photoshop (Figure 3.16) is another perennial favorite, capable of converting all sorts of exotic and complex formats.

- **Music and audio.** MP3 is the clear winner here, and just about all audio programs recognize it on all computers. Microsoft's WAV format is also a good pick for files that will stay in the Windows world, but a WAV file is huge, compared with the same music saved as an MP3 file. To convert other formats and copy tracks from an audio CD to MP3, I use Musicmatch Jukebox Plus (www.musicmatch.com), shown in Figure 3.17; the basic version is available for free!

- **Video.** Your best choices for storing digital video on the PC are Microsoft AVI format or MPEG format—personally, I prefer

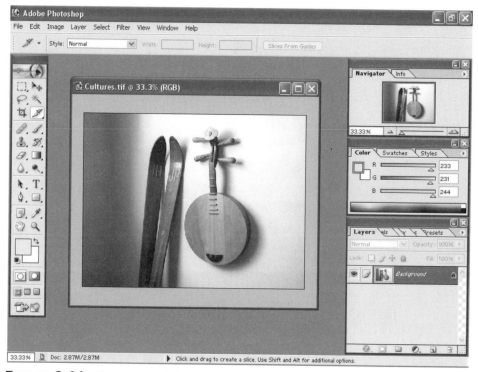

FIGURE 3.16 If you use Adobe Photoshop, put it to work when converting formats.

MPEG format, which is widely used on the Web and is almost always recognized by video editors running on all different types of computers. (If you're recording digital video for use on a Macintosh, consider QuickTime MOV format.) Adobe Premiere makes a great conversion utility, because it's available on both the PC and Macintosh. Figure 3.18 illustrates the familiar face of Premiere.

tip One type of video conversion that you may need to make from time to time is a switch from the PAL video standard used in Europe to the NTSC video standard used in the United States. ArcSoft ShowBiz comes in handy as a conversion tool for this job. For more information on this conversion, see Chapter 12, "Making Movies with ArcSoft ShowBiz," where I give you all the details on ShowBiz.

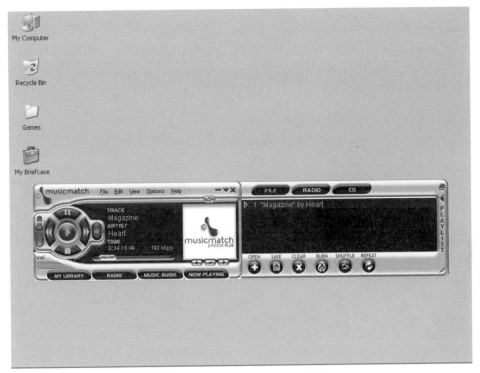

FIGURE 3.17 Musicmatch Jukebox Plus is a great MP3 conversion tool.

- **Documents.** Microsoft Office (Figure 3.19) is the obvious choice here—it's available on both the PC and Mac, and it includes a dizzying array of options for importing, converting, and exporting all sorts of document formats. Microsoft has a passion for preserving older Office document formats, which bodes well for your Word documents in a decade or two.

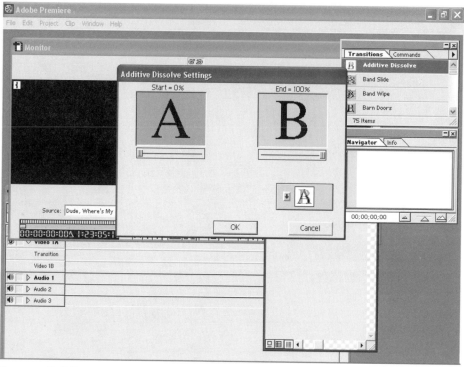

FIGURE 3.18   Adobe Premiere is available for both the PC and the Macintosh.

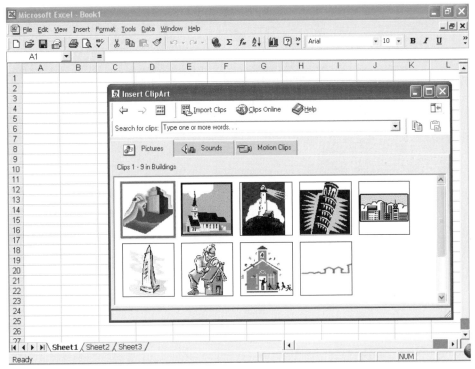

FIGURE 3.19   If you're an Office owner, you already have a great document conversion tool.

Organizing Files

Along with conversion, I strongly recommend that you take a few minutes to plan the arrangement of a data CD before you record it. Sure, you can lump 50 files in the root directory of a DVD and read them without a problem...but will it be easy to find what you're *looking for* in six months?

With convenience and ease of use in mind, here are a few guidelines you can follow:

- **Use folders to store your files.** Grouping your files in folders makes them easier to locate, and some types of discs have a limit to the number of files you can place in the root directory of the disc.

- **Make your names as descriptive as possible.** For example, a file named "State Tax Reports 2001" is easier to locate and easier to comprehend than a file named "STXR2001." (Luckily, Windows allows you to use long filenames.)

- **Organize your files according to a "key."** If you've ever created a database, you know what a key is—it's the data that you use to search your information (whether it be a person's last name or an order number). Arranging files in folders on your disc using a key makes it much easier to locate a specific file in years to come. For example, your key could be the type of file, and all JPEG images would be grouped together. By adding another key—for example, the subject of the image—you can create a folder just for images of your family in JPEG format. (You'll know precisely where all your pictures of Uncle Milton are stored: They're in the folder called "JPEG Family Shots.")

- **Keep things straight with file cataloging software.** At the simplest level, you can use the Windows Search feature to look for a specific filename on your discs—however, there are also shareware and commercial disk cataloging programs that can search for a file through multiple discs and hard drives (even the ones that aren't loaded). Also, you can use a program such as Media Center Plus (again, from Jasc Software at www.jasc.com)—this great utility creates albums of all your multimedia files and allows you to search visually through thumbnails. Figure 3.20 illustrates a thumbnail album from one of my discs. Media Center Plus can also catalog video clips, sounds, and MP3 files, and you can play them back within the program to check out their contents.

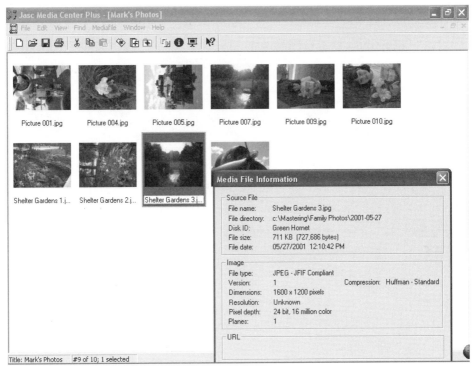

FIGURE 3.20 With Media Center Plus, you can catalog all of your images, sounds, songs, and video.

Summary

In this chapter, I covered the preparations you should make before you launch your DVD recording software. You learned how to optimize the performance of your hard drive by defragmenting and how to add extra memory to speed up your entire system. I discussed how you can delete programs safely to open up additional free space on your hard drive and how to disable the Windows Task Scheduler to prevent interruptions while you're recording. Finally, you learned how to organize files on your discs and how to convert those files to other formats, if necessary.

4

Burning Discs with HP RecordNow

In This Chapter

My second car was a 1979 Ford Pinto hatchback—it was the definition of a basic automobile, with no luxury items besides a cup holder and an AM radio, but I loved that vehicle. (And no, it never showed signs of exploding.) Why do I have such fond memories of my Pinto? To wit, it never broke down, it delivered great gas mileage, and it was easy to drive. Also, I got it for practically nothing.

Those same qualities will endear you to Hewlett-Packard's RecordNow program—well, all except the great gas mileage, anyway. RecordNow is included with the DVD-Writer drive, so you can use your new recorder the minute you install it. RecordNow is very easy to operate, as well. Although RecordNow doesn't deliver the exotic formats of programs such as Easy CD Creator, it does produce most of the basic discs you're likely to need: data DVDs and CDs, audio CDs, and copies of existing discs. I've found RecordNow to be as reliable as my Pinto, and I use this great program every day.

n this chapter, I'll use RecordNow to demonstrate how you can burn each of these standard disc types, and I'll also provide two step-by-step projects: one to record an audio CD from MP3 files and one to burn a disc to carry data while I'm traveling on the road.

Introducing HP RecordNow

You can run RecordNow directly from the Windows Start menu—click Start | All Programs | HP RecordNow | RecordNow—but the easiest way to run the program is simply to load a blank CD, DVD+R, or DVD+RW disc into the DVD-Writer, which automatically displays the Create dialog you see in Figure 4.1. Click HP RecordNow, and click OK when the program recommends a CD-R disc; the RecordNow window appears, as shown in Figure 4.2.

tip **Are you running Windows XP? If so, you may find that Windows displays that irritating "What Should I Do?" dialog (shown in Figure 4.3), along with the Create dialog. If so, you can easily "teach" Windows not to repeat this annoying performance: Click the Take no action entry from the list and click the Always do the selected action checkbox to enable it, then click OK.**

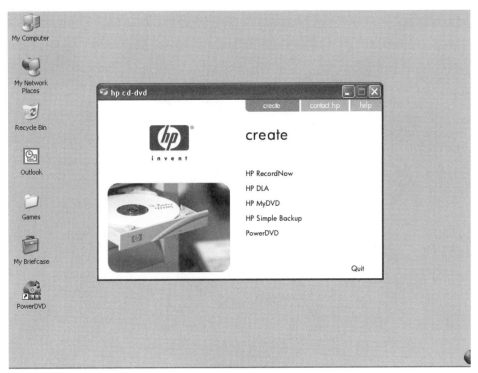

FIGURE 4.1   Loading a blank disc displays the Create window.

As you can see, RecordNow—like my trusty Pinto—includes just what you need, and the program couldn't be easier to use. Because the program doesn't have a traditional menu bar, however, there are two features that may not be immediately apparent:

- The RecordNow Help system appears when you press F1 or when you click the Help button at the top of the window. You can find details on everything the program can do (as well as a direct link to the HP RecordNow support site on the Web) from this menu.
- Click Options to set up general program preferences, advanced recording options, and links to the Internet audio CD database. I'll be discussing many of these options later in this chapter, but make note of the Options menu now, in case you need to change your RecordNow configuration.

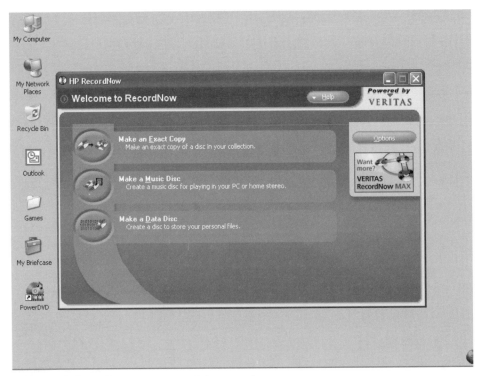

FIGURE 4.2   The main RecordNow window.

tip One option that you'll either love or hate is the "Eject drive trays automatically" checkbox; by default, RecordNow will eject the drive tray when a disc needs to be loaded. This is a boon for the inexperienced users who have multiple drives, but you may want to disable this option if your PC has a door or cover that hides the recorder from view! (There's nothing quite like the sound of your DVD recorder's drive tray making contact with another surface.)

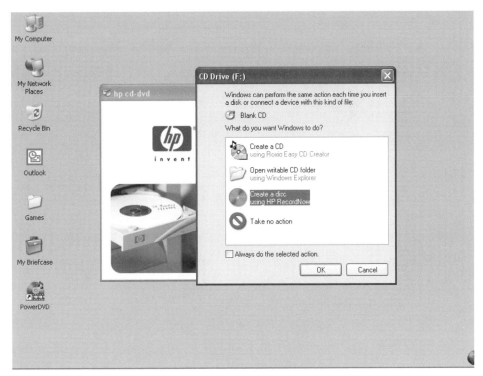

FIGURE 4.3   Windows XP tries to make things easier...no thanks.

Putting Computer Files on a Disc

If your hard drive is running out of room or you want to archive files from your computer for safekeeping, look no further than RecordNow: The program can create data discs using CD-R, CD-RW, and DVD+RW media. (If you've bought an HP DVD-Writer DVD200-series drive, you can also use DVD+R media.)

✔ **Follow these steps to create a data disc:**

1. Click Make a Data Disc; RecordNow displays a wizard window prompting you to insert a blank disc (as shown in Figure 4.4). If you ran the program by loading a blank disc in the first place, continue by clicking Next; if you ran RecordNow from the Start menu, load a blank disc in your drive and click Next.

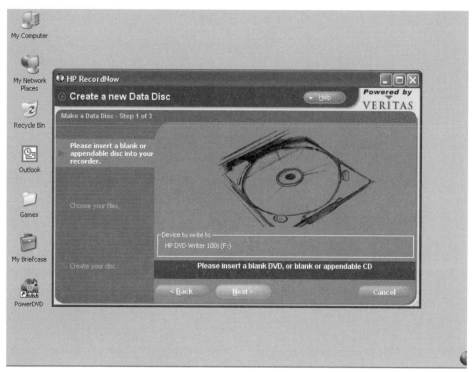

FIGURE 4.4   Time to load a blank CD or DVD, if necessary.

2. Time to select the files you want to add to your data disc: Record-
    Now supports "drag-and-drop" file selection from the Windows
    Explorer, so you can select files and folders and drag them to the
    wizard window shown in Figure 4.5. Alternately, you can click Add
    Files and Folders—RecordNow displays the familiar Windows file
    selection dialog, where you can navigate to the files and folders you
    want to add, select them, and click Add. Whichever method you
    use, RecordNow displays the files you've added in the window and
    updates the amount of space remaining on the disc. To rename a file
    or folder, click the name once, press F2, and type the new name.
    When you've added all of the files you want to add to your data
    disc, click Next to continue.

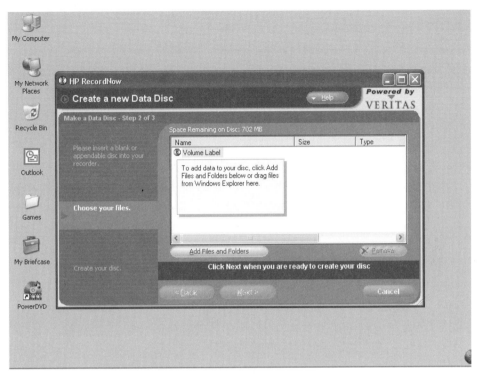

FIGURE 4.5   Drag and drop files on this window, or use the Windows file selection dialog.

tip **To change the volume label displayed for the disc under Windows—commonly called the "name" of a data CD—click on the words *Volume Label* next to the disc icon, press F2, and type a new label, then press Enter to save it.**

3. RecordNow displays a warning dialog, explaining that recording errors can occur if you start new applications while the disc is burned. Generally, this is a good idea if your recorder doesn't offer burn-proof recording—however, because all DVD-Writers support burn-proof recording, you can effectively ignore this warning if you need to take care of other tasks. (To disable this warning, click the Don't show me this dialog again checkbox.) Click OK to continue, and sit back and relax while the recording process begins.

tip By default, RecordNow verifies the data it's written to the disc by comparing the files on the disc with the files on your hard drive—a good way to guard against media defects. Because this can take half as long as the actual recording, you may want to disable verification: Click Options on the RecordNow main window and click the Verify the data written to the disc after write checkbox to disable it, then click OK to save your change.

4. Once the data has been successfully written and verified, Record-Now ejects the disc and displays the dialog you see in Figure 4.6; you can click Make Another to create another copy of the same disc or click Done to return to the RecordNow main window.

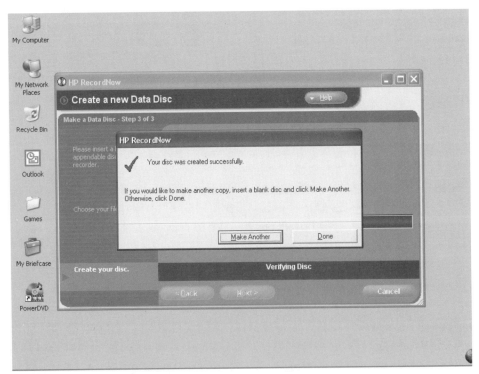

FIGURE 4.6   Another successful burn—I love that Big Green Check!

Recording Your Music

RecordNow makes a great tool for burning audio CDs for use in your home and car audio CD players. The program can create three different types of audio CDs:

- A standard audio CD using MP3 and WAV format digital audio tracks from your hard drive
- A standard audio CD using individual tracks you've copied from existing CDs (a process popularly called *ripping*)
- A special MP3 disc for use in MP3 audio players; these discs can't be used in a standard audio CD player, but they can contain much more music than a standard audio CD. This type of disc is actually not an audio CD at all; instead, it's a data CD that can be read by the improved electronics inside today's MP3 audio players. In effect, the player reads the music files just as your computer does when it plays MP3 audio directly from your hard drive!

tip **If you choose one of the first two types (those that create a standard audio CD), RecordNow will first convert the audio files into WAV format before they're recorded, so avoid burning audio CDs when your drive is very low on free space.**

To burn an audio CD, run RecordNow and click Make a Music Disc. RecordNow prompts you for the type of disc you want to create, as illustrated in Figure 4.7. Click the desired type, and follow the corresponding steps in the next section.

caution **Unless your audio CD player specifically supports CD-RW discs, you should always use CD-R discs to record standard audio CDs! Most audio CD players can't read an audio CD recorded using a CD-RW disc. (Note that MP3 music discs are the exception to the rule, because virtually all MP3 audio players can read CD-RW discs.)**

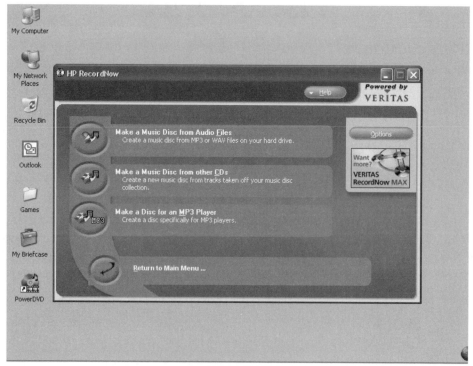

FIGURE 4.7   Selecting a music disc.

Recording with Digital Audio Files

✔ **Follow these steps if you clicked Make a Music Disc from Audio Files or Make a Disc for an MP3 Player:**

1. RecordNow displays the same wizard screen you saw in Figure 4.4, prompting you to insert a blank disc; if necessary, load a blank disc in your drive and click Next.

2. To search for MP3 and WAV digital audio files on your system, click Yes at the prompt; RecordNow will find all digital audio files in these two formats on your hard drive. If you have an Internet connection

active, the program will also attempt to identify the tracks, using the CDDB Internet database.

tip **Are you a game player? If you choose to search for files, be prepared for a long wait—and several dozen MP3 and WAV files that you probably don't want to record to an audio CD. (Most of the PC games written today use MP3 audio for background music and WAV files for sound effects.) In fact, Windows itself ships with a number of WAV files that will also show up. The moral of the story? If you know what audio files you want to record and where they are, click No at the search prompt and skip to step 3.**

3. RecordNow displays the layout wizard screen you see in Figure 4.8. If you searched for audio files, the files that were found are listed in the Music on System list at the left; you can click a filename to select it in this list and click Add to add the song to your music CD layout. To add specific files from your hard drive, you can drag and drop files from the Windows Explorer; alternately, you can click Browse to display the Windows file selection dialog, where you can navigate to the audio files, select them, and click Add. The files you add appear in the list on the right, and RecordNow updates the amount of time remaining. When you've finished adding tracks, click Next to begin recording.

tip **You can rearrange the order of the tracks in the layout by dragging a filename to the desired location. To preview an audio file, click the desired filename to select it and click the speaker icon under the Music on System list.**

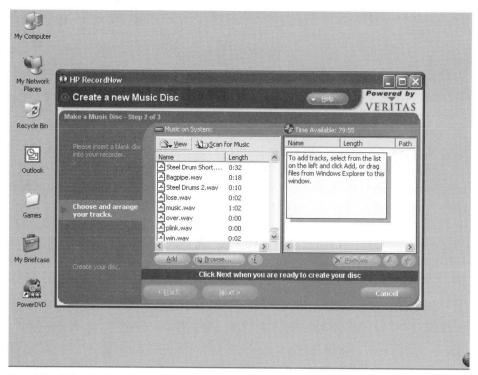

FIGURE 4.8   Adding MP3 and WAV digital audio to a CD.

Recording Tracks from Existing Audio CDs

✔ **Follow these steps if you clicked Make a Music Disc from other CDs:**

1. RecordNow automatically ejects the disc tray and displays the wizard screen shown in Figure 4.9, prompting you to insert a music CD. Load the first disc containing the music you want to record and click Next.

2. RecordNow displays the layout wizard screen—the tracks on the CD you loaded are listed in the Music CD Tracks list at the left; click a track name to select it in this list and click Add to add the song to

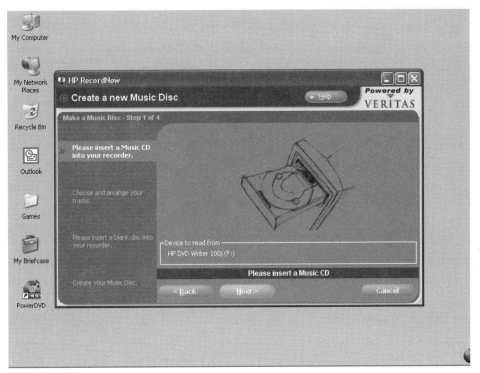

FIGURE 4.9   Load your music and prepare to rip.

your music CD layout. Remember, to preview a track, click the desired track name to select it and click the speaker icon under the Music CD Tracks list (click the button again to exit preview mode). The files you add appear in the list on the right, and RecordNow updates the amount of time remaining.

3. When you've finished adding tracks from the current audio CD, click Next CD to load the next source disc. RecordNow temporarily copies the tracks you selected to your hard drive.

4. When you've finished adding tracks, click Finish. RecordNow prompts you to load a blank CD to begin recording.

Copying a Disc

✔ **You can also use RecordNow to make an exact duplicate of an existing data or audio CD. Follow these steps:**

1. Run RecordNow and click Make an Exact Copy; RecordNow automatically ejects the disc tray and prompts you to load the disc you want to duplicate.

2. Load the source disc and click Next.

3. RecordNow displays a progress bar as it copies the contents of the source disc to a temporary image file on your hard drive. Once the original disc has been copied to the computer's hard drive, the program ejects the tray and prompts you to load a blank CD to begin recording. If you use a second CD-ROM drive to read the source disc while your DVD-Writer concentrates on recording, no disc swapping is necessary: Simply congratulate yourself for your foresight and consider yourself a "Power Ripper."

tip **Don't attempt to copy a "copy-protected" CD or DVD disc with this program (such as a DVD movie or a game CD-ROM). Naturally, RecordNow is not meant for duplicating these discs, and the result will be a shiny coaster.**

Project: **Recording a "Road Trip" Audio CD**

What could be better on a long drive than a collection of your disco favorites? (Don't answer that...there's always the Partridge Family.) Anyway, in this project, we'll burn a "Road Trip" audio CD, using tracks taken from several different existing audio CDs: The resulting disc can be played in any car or home audio CD player.

Requirements

- A collection of audio CDs
- Blank CD-R disc

✔ **Follow these steps to create your compilation audio CD:**

1. Run RecordNow and click Make a Music Disc, then click Make a Music Disc from other CDs.

2. RecordNow automatically ejects the disc tray and prompts you to load the first of the existing audio CDs. Load the first disc containing the music you want to record and click Next.

3. Figure 4.10 illustrates the tracks from my first source CD; let's add *A Fifth of Beethoven* and *Play That Funky Music*. Click *A Fifth of Beethoven*, hold down the Ctrl key, and click *Play That Funky Music* to select both tracks, then click Add. Note that RecordNow indicates we have 66 minutes and 5 seconds of time remaining on this disc.

4. Click Next CD to load the next source disc; RecordNow takes a few seconds to copy those first two tracks temporarily to your hard drive, then ejects the drive tray. Load the next disc and click OK.

5. Repeat steps 3 and 4 until you've added all of the desired tracks; for example, Figure 4.11 illustrates my completed disco audio CD layout, using selected tracks from three different audio CDs. However, I've decided that I'd rather start the disc with *YMCA*—I'll click on the name *YMCA* in the track list and drag it to the first position in the list.

6. Time to burn this disco inferno! Click Finish. RecordNow prompts you to load a blank CD, and the recording begins.

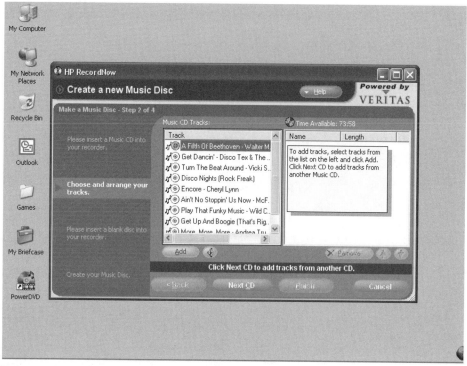

FIGURE 4.10   Selecting the perfect disco tracks.

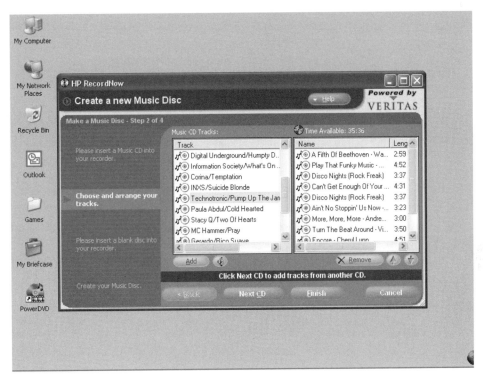

FIGURE 4.11 A completed disco disc layout.

Project: **Creating a "Briefcase Backup" DVD**

Next, I'll record what I like to call a *Briefcase Backup* (if you've read other books I've written, you'll probably recognize the term). A Briefcase Backup is a disc that you record to augment your laptop's hard drive; for example, when I travel, I store all of the data I'll need (including things such as PowerPoint presentations, Excel worksheets, books in PDF format, and digital video samples) on a CD-RW disc. (I could do the same thing with a DVD+RW disc, but I don't need to haul around that much data yet!) A Briefcase Backup has a number of advantages for the road warrior:

- Most laptops (especially older models) have smaller hard drives—rather than use up that valuable space, you can store 700 MB of your traveling data on a disc instead.
- In case of a damaged or lost notebook, you still have a backup if you carry the disc in your luggage.
- If you need to transfer that data to a client's PC, there are no silly cables or network cards needed; you can load the disc in any PC with the proper drive.

In this case, I'll record all of the files necessary to recreate this entire book to a Briefcase Backup—that includes Word documents, templates, and every single figure, all of which fit comfortably on a single DVD+RW disc!

Requirements

- Blank CD-R, CD-RW, DVD+R (if you're using a DVD-Writer DVD200 drive), or DVD+RW disc

✔ **Follow these steps to record data:**

1. Run RecordNow and click Make a Data Disc, then load the blank disc (if necessary) and click Next to continue.

2. I'll use drag-and-drop file selection from Windows Explorer; Figure 4.12 shows Windows Explorer open, displaying the folder with the files I need to record. Select the desired files—remember, you can hold down Ctrl while you click to select multiple files, then click and drag the files to the wizard window. To rename a file or folder, click the name once, press F2, and type the new name.

3. Before I record a Briefcase Backup disc, I usually name it with a description of the contents; in this case, we'll change the volume label to DVDREC. Click on the words Volume Label next to the disc icon, press F2, type DVDREC, and press Enter. That completes the data layout, so click Next to continue.

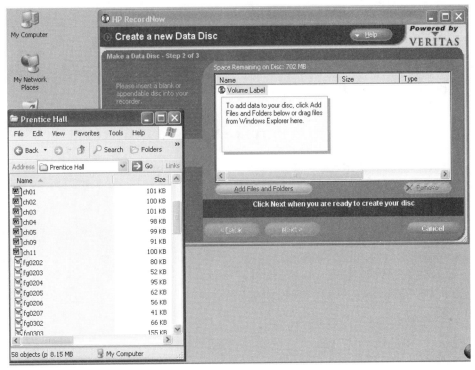

FIGURE 4.12 Ready to drag and drop files on the RecordNow wizard window.

4. If you haven't disabled the warning dialog, click OK to continue. RecordNow begins burning our disc.

5. When the verification completes successfully, our new Briefcase Backup disc is ejected from the drive. RecordNow prompts you for additional copies of the same disc. Because we need only one copy, click Done to return to the RecordNow main window.

Project: **Copying a Backup Data CD for Archiving**

To close this chapter, let's prepare your company's data (or the data from your home office computer) against the worst by creating a second copy of your hard drive backup disc for storage off site. There are numerous companies that can perform this operation for you and store your archival copy—for a hefty fee, of course—but by using your DVD or CD media and RecordNow, you can make a copy and store it in your safety deposit box at your bank. Total cost? Probably less than $5—and, if you use rewriteable media, you can continue to update your archival copy on a regular basis!

Requirements

- The original disc
- A blank DVD+RW or CD-RW disc

✔ Follow these steps to record data:

1. Run RecordNow and click Make an Exact Copy; RecordNow prompts you to load the original data disc, as shown in Figure 4.13.

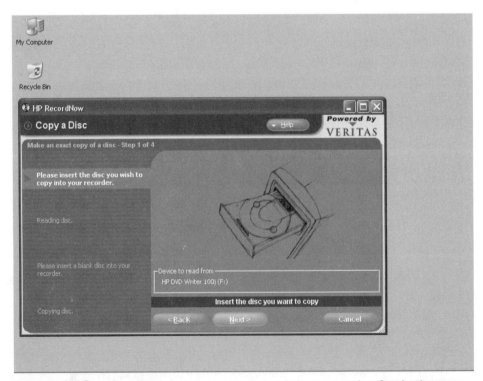

FIGURE 4.13 RecordNow prompts you to load the source disc for duplication.

2. Load the source disc—in this case, the backup data disc—and click Next to close the drive door and begin the duplication.

3. If you're using two drives (your DVD recorder and a CD-ROM or DVD-ROM read-only drive for the source) to copy, just load the blank disc in the recorder and grab yourself a cup of coffee or refill your soda—there's no need to swap discs, so your job is finished.

4. If you're using only your DVD recorder, sit back and relax while RecordNow copies the contents of the backup data disc to your hard drive (Figure 4.14)—this image will be used as the source when you load the blank disc into your recorder.

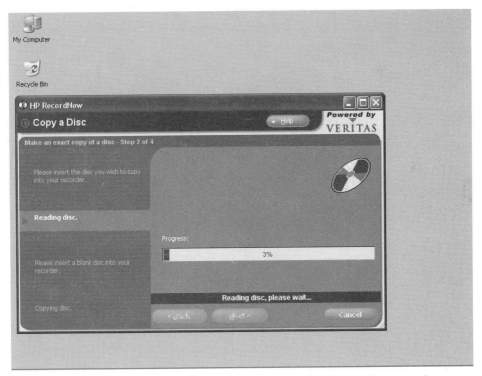

FIGURE 4.14 When copying with only one drive, the data is first copied to your hard drive.

5. If you're using one drive, the program ejects the tray, once the original backup data disc has been copied to the computer's hard drive, and prompts you to load a blank disc to begin recording (as shown in Figure 4.15).

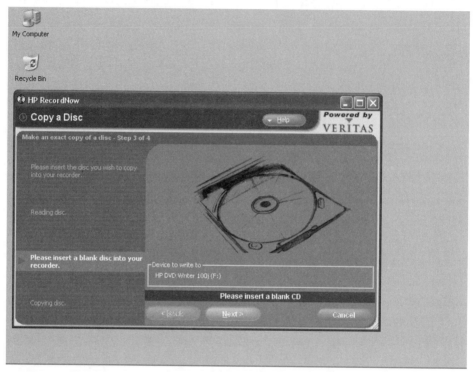

FIGURE 4.15 Load a blank disc, and RecordNow does the rest!

Summary

In this chapter, you learned how to create the "bread-and-butter basics" of CD and DVD recording: data discs (created from files and folders on your hard drive), audio CDs (burned from digital audio files on your hard drive or tracks ripped from existing music discs), and MP3 discs (data discs that contain digital audio files for use in MP3 audio players).

5

Drag-and-Drop Recording with HP DLA

In This Chapter

✓ Formatting and writing DLA discs

✓ Ejecting a DLA disc

✓ Adding files to an existing DLA disc

✓ Project: Creating a Genealogy Archive Disc

✓ Project: Recording and Finalizing a PowerPoint Presentation Disc

UDF (short for Universal Disc Format) recording, also called packet writing, is practically the perfect method of burning data files over an extended period of time. You can copy files to or delete them from a CD-R (or, if you're using the HP-Writer DVD200-series drive, a DVD+R) UDF disc as you would a floppy drive, a ZIP drive, or a hard drive. In fact, if you can save a file using drag-and-drop, Windows Explorer, or any other Windows application, everything is transparent—you don't have to worry about running a separate program, creating a layout, or recording a session.

Naturally, UDF can team up with rewriteable CD-RW and DVD+RW discs, too—most computers with a DVD-ROM drive running Windows or Mac OS can read a DVD+RW UDF disc, and any computer with a MultiRead CD-ROM drive can read a CD-RW UDF disc. Because both of these media types can be reformatted and used again, your DVD+RW drive becomes what many computer owners consider the "perfect" removable media drive!

The Hewlett-Packard DVD-Writer drive comes complete with HP DLA (short for Drive Letter Access), a complete UDF formatting and burning application; I'll use this program while covering how to format a UDF disc, how to copy files to it, and how to eject it properly when you're done. Our three projects for this chapter illustrate how you can create a removable disc for carrying files between your work and your home computers, how to record a DVD genealogy archive disc that you can update whenever you like, and how to burn a PowerPoint presentation disc that's finalized for use in any PC CD-ROM drive.

Formatting a DLA Disc

Like any floppy disk, ZIP disk, or hard drive, a UDF disc must first be formatted before you can write to it. You can do this from the Toolkit, which is available from the Windows Start menu—click Start | All Programs | HP DLA | DLA Toolkit.

Of course, the easier this task is, the better—therefore, you can also simply load a blank CD or DVD disc into the DVD-Writer, which automatically displays the familiar Create dialog. Click HP DLA, and click OK when the program recommends a CD-R or DVD+RW disc— I actually use UDF more often with rewriteable media, such as CD-RW or DVD+RW discs.

Figure 5.1 illustrates the Toolkit window that appears regardless of which method you choose to display the program.

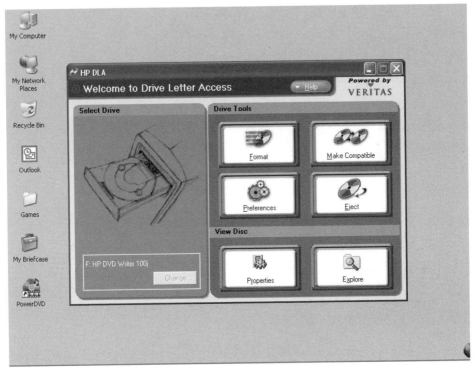

FIGURE 5.1   The main DLA Toolkit window.

✔ **Follow these steps to format a disc for use with HP DLA:**

1. Click Format; the Toolkit displays a wizard window prompting you to insert a blank disc (as shown in Figure 5.2). If you ran the program by loading a blank disc in the first place, continue by clicking Next; if you ran DLA from the Start menu, load a blank disc in your drive and click Next.

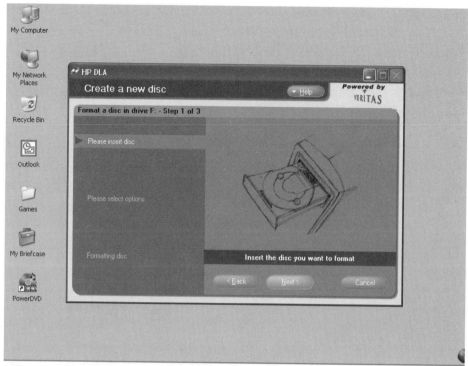

FIGURE 5.2   DLA prompts you to load a blank disc.

2. Next, it's time to configure the disc. First, click in the Volume Label field and type a name for the disc; this is the name that will appear with the DLA drive in Windows Explorer.

3. DLA can use either of two formatting modes: You can choose between a *quick format* and a *full format*. As the name implies, the Quick format is much faster (taking less than a minute), but it can be used only in certain cases: if you're formatting a CD-R disc for the first time or if you're reformatting any type of rewriteable media. On the other hand, the Full format must be used if you're formatting a CD-RW or DVD+RW disc for the first time.

> *tip* Generally, I use Full format only when initially formatting CD-RW or DVD+RW discs; however, if you have a rewriteable disc that returns errors when you try to read it (or can't be read at all),

you should always use Full format before using it again. Such errors can occur because of a power failure or lockup while your PC is writing to the disc.

4. Finally, click the Enable compression checkbox if you'd like to increase the capacity of the disc (generally by a ratio of 2:1, depending on the type of data you're storing)—however, enabling compression means that other computers will need HP DLA installed before they can read your disc. Without compression, any computer with a compatible drive that supports UDF discs will be able to read the disc—this includes all current versions of Windows, as well as Mac OS 9.x, Mac OS X, and most "flavors" of UNIX and Linux. Figure 5.3 illustrates the settings I'm using for a DVD+RW disc. Click Next to continue.

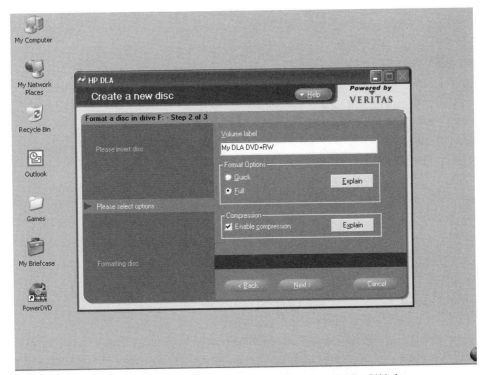

FIGURE 5.3   Configuring the formatting settings for a DVD+RW disc.

5. The Toolkit program displays a progress bar and an approximate time remaining to complete the format; once the disc has been successfully formatted, you're returned to the Toolkit main window. To display the capacity, compression settings, and volume name, click Properties to display the Properties dialog (Figure 5.4). Click Close to return to the Toolkit window.

tip **If you've enabled compression on your DLA disc, you may be wondering where all that extra space comes in—because the compression is applied when the files are written, DLA can display only the used space and an *approximation* of the free space remaining. It pays to keep an eye on the free space for a UDF disc after writing to it, because Windows can only estimate the amount of space left.**

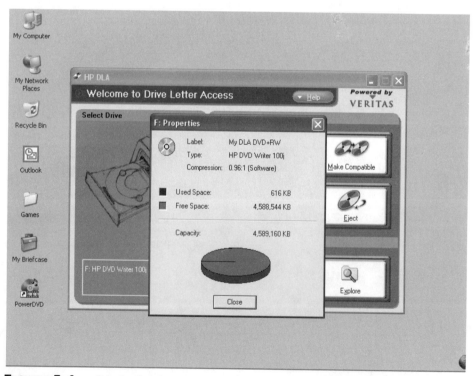

FIGURE 5.4   Displaying the properties for a newly formatted HP DLA disc.

Writing Files to a DLA Disc

Don't expect to find complex procedures in this section! DLA operates *entirely* in the background—nearly invisibly—while you use your new CD or DVD disc. You can read, copy, and move files to your HP DLA disc using the same standard Windows menu commands, keyboard shortcuts, and drag-and-drop mouse operations that you're already using—you can also delete files from the DLA disc (although you can't reclaim the space occupied by the deleted files if you're using CD-R or DVD+R media).

tip **DLA discs are data-only: you can't create an audio CD using HP DLA, and they're not suitable for use in audio CD players. However, you can use a DLA disc to store MP3 files for later. (When you have the chance, you can copy the digital audio files to your hard drive and burn them to a music disc, using HP RecordNow.)**

Your applications can open and save data to documents on a DLA disc, just as they would with a hard drive—you can also use the Send To option that appears when you right-click on a file or folder to copy a file to your DLA disc. Even if you use Windows Explorer or another file management program, there are no settings to change or adjustments to make to use a DLA disc; it's all taken care of by the HP DLA program, and that's all there is to it! Figure 5.5 illustrates how the DLA disc that I formatted in the last section looks in Windows XP's My Computer window.

✔ **For example, if you have a file saved on your Windows desktop, follow these steps to copy it to your DLA disc:**

1. Double-click My Computer to open it and display your drive icons.

2. To copy the file into a folder on your DLA disc, double-click the icon for your DLA disc and double-click the desired folder to open it.

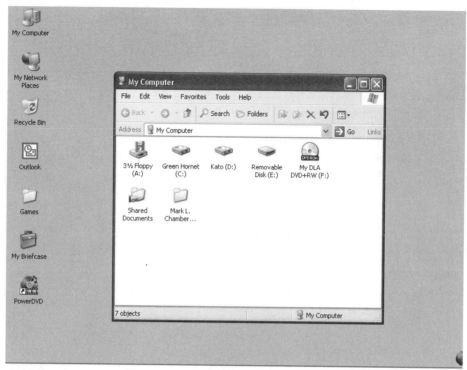

FIGURE 5.5   To Windows XP, my DLA disc is just another removable disc.

3. Click the file you want to copy and continue to hold down the mouse button. Drag the file to the DLA folder window to copy it.

Ejecting a DLA Disc

In most cases, ejecting a DLA disc is no different from ejecting a regular read-only CD or DVD. You can:

*   Press the Eject button on your drive.
*   Right-click on the disc icon in the My Computer window.
*   Run the HP DLA Toolkit application and click the Eject button.

Note that DLA may require a few seconds to prepare the disc before it's ejected (depending on whether the drive is still writing data).

tip **If you're planning on reading the disc on a computer that's not running a UDF-compatible operating system (for most of us, this means Windows 95 and earlier versions or Mac OS 8.5 and earlier versions), you should make your UDF disc** *compatible* **first—I'll discuss this feature a little later in this chapter.**

Recording Additional Files to a DLA Disc

"Can I add more files to a DLA disc after I've ejected it?" You bet! In fact, this is yet another reason why I call creating discs with UDF programs such as HP DLA "the perfect method of burning data files." As long as you don't make your DLA disc compatible (as I show you in the next section) and your disc has free space remaining to fill, simply load your disc back in the drive, and DLA will automatically recognize it.

In fact, with HP DLA, this process of "mounting" a UDF disc operates totally "behind the scenes"—unlike some other UDF programs, HP DLA doesn't display a dialog or any indicator that it's working, so you just load the disc and go.

Making a DLA Disc Compatible

If there's any drawback at all to a DLA disc, it's the fact that you can't read the contents of that disc in two situations:

1. If you've used compression on the disc and the computer you're using doesn't have HP DLA installed

2. If the computer doesn't support UDF discs (remember, this means Windows 95 and earlier versions or Mac OS 8.5 and earlier versions)

To be honest, there's not much you can do in the first situation—if you've chosen compression, HP DLA *must* be installed on the PC, or you won't be able to read your DLA disc, period. However, there is hope in the second situation: If you want to make a DLA disc readable on older versions of Windows and Mac OS, you can make the disc *compatible*, effectively turning your UDF disc into an *ISO 9660* disc. (ISO 9660 is a universal disc format that's recognized by just about every computer with a CD-ROM drive on the planet.)

To make a disc compatible, you must be using a CD-R disc, and the disc must be formatted *without* compression—also, making a disc compatible will close (or finalize) the disc, which prevents you from writing additional data to it. Therefore, never make a disc compatible until you've finished writing all of the files you want to the disc.

✔ **Follow these steps to make a DLA CD-R disc compatible:**

1. Double-click on My Computer and right-click on the DLA disc icon; select Make Compatible from the pop-up menu.

2. HP DLA displays the confirmation dialog you see in Figure 5.6. Click Start to begin the compatibility processing.

3. Once the compatibility procedure has completed, HP DLA displays a status dialog, and you can eject the disc.

Note that making a disc compatible will shorten long file names, and directories nested more than 30 folders deep will not be readable. Compatible discs cannot be read by Windows when you're using DOS mode. If you're using CD-RW or DVD+RW discs, you can quick-format them and reuse the discs after they've been made compatible.

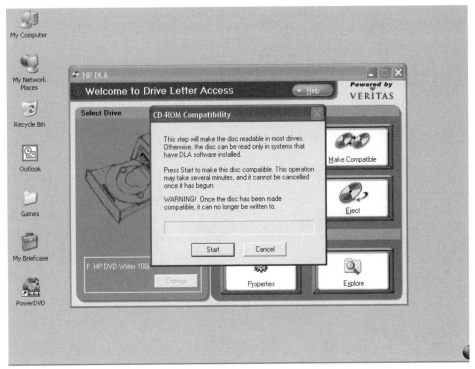

FIGURE 5.6 Are you absolutely, positively sure you want this disc compatible?

Project: **Recording a DVD Genealogy Archive Disc**

I can't think of a better use for a DLA disc than a genealogy archive— your valuable data is stored safely, but you can update the disc with additional information that you turn up with continuing research. (Come to think of it, the same is true for your Quicken data files and TurboTax returns for the last five years...but we'll stick with the gene- alogy project, because I thought of it first.) With compression, you may be able to store 5 or 6 GB of data on one disc; that's enough space for reunion videos, a library of voices in WAV format, and a high-reso- lution Windows bitmap image of every member of your clan!

Because you won't be distributing this disc to others, you can safely use compression and a DVD+RW disc. We'll assume that your blank DVD+RW disc is unformatted.

Requirements

- A collection of genealogical data files
- A blank DVD+RW disc

✔ **Follow these steps to create your genealogical archive:**

1. Run the DLA Toolkit and click Format. Load a blank disc in your drive and click Next.

2. Click in the Volume Label field and type the name Family Data for the disc. Because this is the first time you've used this DVD+RW disc, click Full format. Click the Enable compression checkbox to turn on software compression. Click Next to continue.

3. Once the formatting process has finished, you're ready to copy your data files! Click Explore from the Toolkit window to display the contents of the Family Data disc—naturally, it's currently empty. Double-click the My Computer icon on your desktop and navigate to the location of the files you want to store, then select them and drag them to the Family Data window (Figure 5.7).

4. Once you've finished copying files to the archive disc, you can eject it and store it for later use; when you're ready to copy additional files to the DLA disc, just load it in the DVD-Writer drive again, and HP DLA will recognize it. For example, if you want to save a copy of a Word document that you've just written, click File and choose Save As, then navigate to your DLA disc and click Save.

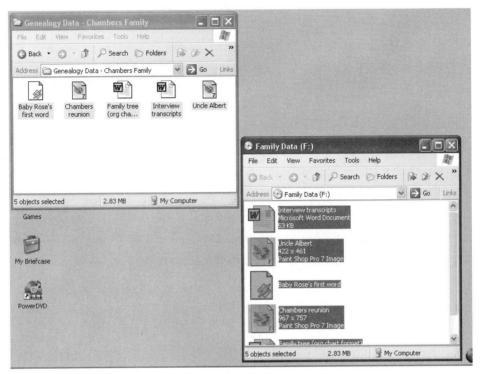

FIGURE 5.7   Storing family data optically for generations to come.

Project: **Recording a "Working Copy" UDF Disc**

Now let's consider what I call a "working copy" disc: It's a disc with your current project and data files that you carry with you between your computer at home and your computer at work. (A working copy disc also does double duty as a simple backup in case one of these two computers goes haywire.) We'll assume that one of your computers uses HP DLA, but the other does not have the program installed; naturally, this means you can record additional files on only one computer.

   In this situation, we want to leave the disc compatible with a computer that's not running HP DLA, and we also don't want to finalize it. This way, you can continue to add (and overwrite) files until you've reached the maximum capacity of a CD-R disc.

Requirements

- Programs and data from your home or office
- A blank CD-R disc

✔ **Follow these steps to record data:**

1. Run the DLA Toolkit and click Format. Load a blank disc in your drive and click Next.

2. Click in the Volume Label field and type the name *Working Copy* for the disc. Click Quick format, and make sure the Enable compression checkbox is turned off. Click Next to continue.

3. Use drag-and-drop or the Send to menu item from the right-click menu to copy files.

4. When you're ready to take the disc to your other computer, eject it. Additional files can be recorded to the disc by simply reloading it in the computer that's running HP DLA.

Project: **Recording and Finalizing a PowerPoint Presentation Disc**

Our third project in this chapter is a little trickier: You need to create a CD-R containing your PowerPoint presentation files that you can send to your company's overseas branch. Unfortunately, not all the files have been assembled yet—you're still waiting for logos and artwork from the marketing department—and you're not sure what type of computer will be reading the disc at the overseas office.

Fear not! HP DLA comes to the rescue again: You can create the presentation disc in segments, saving the files you have on hand until they're all assembled. In addition, you can make the resulting disc compatible, and virtually any computer (including Macs and Linux machines) should be able to read it.

Again, we'll use a blank CD-R disc straight from the spindle.

Requirements

- PowerPoint data files
- A blank CD-R disc

✔ **Follow these steps to record data:**

1. Run the DLA Toolkit and click Format. Load a blank disc in your drive and click Next.

2. Click in the Volume Label field and type the name PowerPoint Project for the disc. Click Quick format (when you're formatting a CD-R DLA disc, you can use this time-saver), and make sure the Enable compression checkbox is turned off—remember, you can't make a disc compatible if it's formatted with software compression! Click Next to continue.

3. Use drag-and-drop or the Send to menu item from the right-click menu to copy the first batch of data files to the new DLA disc, then eject it until the remaining files arrive.

4. Finally got those graphic files from Marketing? Load the Power-Point Project DLA disc in the DVD-Writer drive and copy the updated presentation to your disc; feel free to "delete" older versions of the presentation. (As I mentioned earlier, this won't reclaim the space, but with 700 MB of elbowroom, you can probably fit several iterations of a typical PowerPoint project on a single CD-R DLA disc.)

5. When the disc is complete and ready to send, double-click on the My Computer icon on your desktop and right-click on the Power-Point Project disc icon; select Make Compatible from the pop-up menu.

6. Click Start on the confirmation dialog to begin the compatibility processing.

7. Once the compatibility procedure has completed, eject the disc, and you're ready to send it! (If you like, you might also make a backup copy that you can store; run the HP RecordNow program and use the Make an Exact Copy feature to create an exact duplicate of the disc, as I demonstrated earlier in the book in Chapter 4, "Copying a Disc.")

Summary

We explored the convenience and ease of UDF recording in this chapter—you learned how to use HP DLA to create "removable media" discs for your own use, and I showed you how to make your DLA discs compatible in case you need to swap them with other computers and "foreign" operating systems.

chapter

6

Recording a DVD with Existing Files

In This Chapter

MyDVD is a great example of the new generation of DVD recording software—like Apple's iDVD on the Macintosh side, MyDVD makes it easy for anyone to create a DVD Video disc, complete with basic interactive menus! No longer do you need years of film editing experience, $20,000 of hardware and software, and the eye of a "starving video artist." MyDVD takes video clips directly from your hard drive and allows you to choose things such as backgrounds and buttons from an easy-to-use menu system; there's even a helpful wizard to guide you through the entire process, if you like.

n this chapter, I'll show you all of the features of MyDVD, and we'll create a DVD with video clips of your family's summer vacation taken from your hard drive.

Introducing MyDVD

As you can see from Figure 6.1, there are actually two components to MyDVD: the menu bar and toolbar, which stretch across the top portion of the screen; and the menu editor, which occupies the center of the screen in its own movable dialog. You can resize the menu editor by dragging the lower right corner of the editor window. MyDVD also includes a wizard that we'll use to create a DVD later in the chapter. To run MyDVD from the Windows XP Start menu, click Start | All Programs | MyDVD | MyDVD.

FIGURE 6.1   The MyDVD toolbar and menu editor.

MyDVD can operate in two modes: You can either edit and record existing clips (or clips that you've captured beforehand) or you can record directly from a video source to a DVD. In this chapter, we'll cover the first mode of operation; I'll cover direct-to-DVD recording in Chapter 7, "Direct-to-DVD Recording."

You can even use MyDVD to edit and "remaster" DVDs that you've already created—this is a great way to create an "open-ended" family video, where new scenes can be added and the disc recorded again with the new material (naturally, the rewriteable nature of DVD+RW comes in handy for this kind of project).

As a reminder, both DVD+RW and DVD+R discs are compatible with most DVD players and computer DVD-ROM drives, but there are exceptions; generally, the older the DVD player, the less likely it will be able to read DVD+RW and DVD+R media. Therefore, it's a good idea to verify that you can use DVD+RW or DVD+R discs with the equipment that you'll use to view your DVD movies. The easiest way to do this is to burn a quick test disc and try it out on your DVD player or DVD-ROM drive—you can also check the Web site of the manufacturer of your DVD hardware for a compatibility listing. (If you're interested in using CD-R discs instead, MyDVD can also produce a cDVD disc, which will be compatible with most computer CD-ROM and DVD-ROM drives.)

Using the Wizard

By default, the MyDVD Wizard greets you each time you open the program; Figure 6.2 illustrates the opening Wizard screen. From this screen, you can choose to:

- Record direct-to-DVD (which I will explain in the next chapter)
- Create a new project with existing photos and video clips
- Open an existing project and continue working on it
- Edit an existing DVD

You can also run the program's tutorial from this screen—like many online help systems these days, the tutorial is in HTML format (the same language used to create Web pages), so MyDVD opens your Web browser of choice when you click the Tutorial button. Figure 6.3 illustrates the opening page of the MyDVD tutorial, which you can navigate simply by clicking on the links.

tip **Once you're familiar with using the manual features in MyDVD, you can click the Always show this wizard when MyDVD starts checkbox to disable it, which will take you directly to the MyDVD screen layout I showed you earlier. (This setting can also be toggled on and off from the Preferences dialog, which you can display by clicking File and choosing the Preferences menu item.) Alternately, click Cancel on this opening Wizard screen, and the program will return you to the MyDVD menu bar and menu editor.**

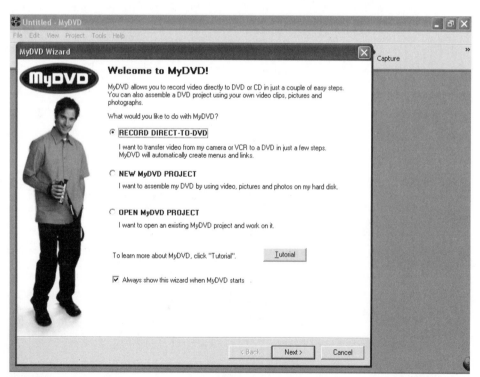

FIGURE 6.2   The Wizard screen is ready to help as soon as you run MyDVD.

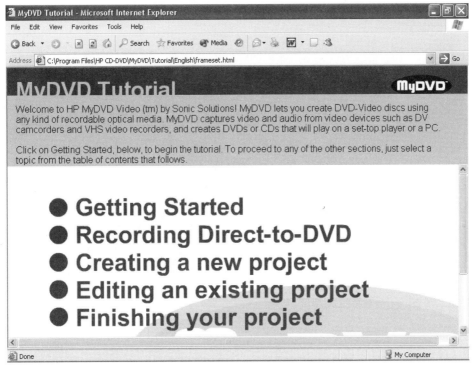

FIGURE 6.3   The MyDVD tutorial uses your Web browser.

I'll describe the Wizard in the project at the end of this chapter—for now, just remember that the process of starting a project can be automated through the Wizard.

Adding a Menu Manually

Like commercial DVD movies, you can create menus with MyDVD that allow branching movement (where you move from menu screen to menu screen by pressing a button on your DVD player's remote control)—each one of these submenu screens can also display buttons that can display photos or run video clips. In this way, you can build a simple menu tree; for example, you might have a Title screen with two submenu buttons. One button takes the viewer to a submenu screen ded-

icated to videos of the family pet, and the other submenu screen would
include videos of your family vacation.

✔ **MyDVD makes it easy to create menus. Follow these
steps to add a menu:**

1. Click the Menu button at the upper right corner of the menu editor window. MyDVD adds the Untitled Menu button you see in
Figure 6.4.

2. Double-click on the Untitled Menu button to display the new submenu screen, as illustrated in Figure 6.5. Note that MyDVD automatically adds two buttons to the bottom of the new screen: the
Home button (which takes you back to the main Title screen of the
menu system) and the Previous button (which takes you back to
the previous screen).

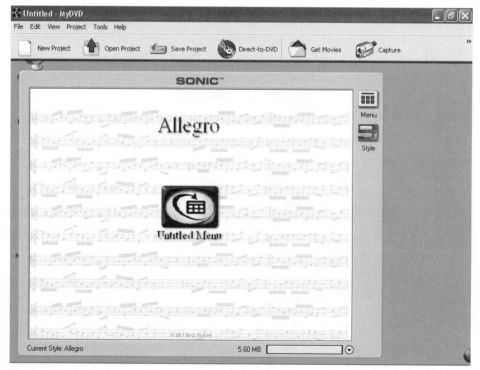

FIGURE 6.4 Adding a menu to a screen adds an Untitled Menu button.

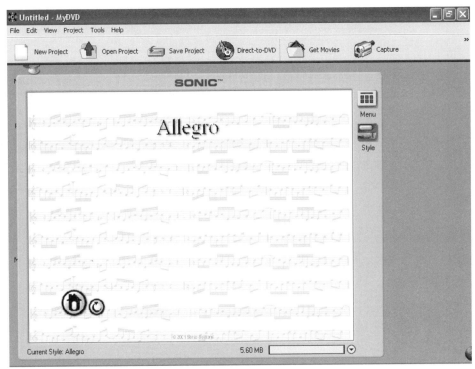

FIGURE 6.5   MyDVD automatically creates a submenu when necessary.

tip   As you add menus, photos, and video clips, you'll note that the bar graph at the lower right corner of the menu editor will indicate how much space the project will occupy on a disc. If you're designing a disc with several large clips, it's a good idea to keep an eye on this gauge to make sure that your project doesn't exceed the capacity of a blank disc. (You can click on the drop-down button next to the gauge to switch between the 4.7-GB capacity of a DVD and the 650-MB capacity of a CD.)

3.   Next, change the text on the menu button you've just created. Double-click on the Previous button—because we've added only one submenu, this button will take you back to the Title screen, as well. Click on the text "Untitled Menu" under the button you added, and MyDVD opens a text editing box (shown in Figure 6.6). Type the new label for the button and press Enter to save it.

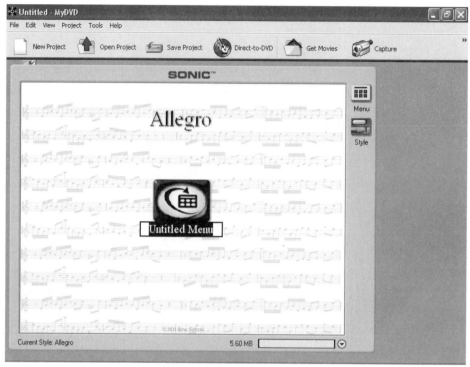

FIGURE 6.6   Editing the text for a menu button.

You can delete the currently displayed menu by right-clicking on the menu background—choose Delete Menu. If the current menu contains one or more buttons (either submenu or movie buttons), MyDVD prompts for confirmation; click OK, and all of the menu and movie buttons (and anything they are linked to) are deleted from your project.

To delete a menu button, right-click on it and choose Delete Button—again, MyDVD prompts you for confirmation, because you'll lose the submenu and anything linked to the button.

Adding Video Clips Manually

✔ **Buttons on a MyDVD menu can also display video clips from your hard drive. Follow these steps to add movies to your DVD project:**

1. Click the Get Movies button on the MyDVD toolbar—the program displays the contents of the My Documents\My Videos folder in the Add movies to menu dialog (Figure 6.7). Navigate to the location where you've saved your movies, select one or more files, and click Open.

 tip **You can also add movies by dragging the files directly from the Windows Explorer onto the menu editor.**

FIGURE 6.7   Selecting video files from the My Videos folder.

2. Neat! MyDVD automatically adds a button that displays the movie, complete with a thumbnail image from the beginning of the video file (as shown in Figure 6.8). The movie's filename is used as the default button; as with a menu button, you can click on the label and type your own text, then press Enter to save it.

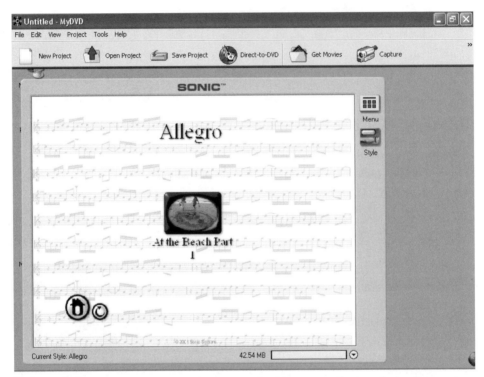

FIGURE 6.8   Our new video clip button looks very professional, don't you think?

3. It's often necessary to *trim* a video clip so that extraneous material at the beginning or ending of the movie doesn't appear in your project; consider this somewhat similar to cropping a digital photo using Paint Shop Pro or Photoshop. (Note that this process *does not* actually remove any frames from your digital video! It only "marks" the beginning and ending of the clip for your MyDVD project.) To trim the clip, double-click on it; MyDVD displays the Trim dialog you see in Figure 6.9, with thumbnail images of the start and ending frame, along with the duration, starting time, and ending time (If you're not familiar with the time notation used in video editing, everything is in hours:minutes:seconds:frames.) A Hollywood film has 24 *frames per second* (or FPS); PAL video has 25 FPS; and NTSC video has 30 FPS (it's actually 29.97, but everyone rounds it upward.)

4. To select the beginning point for your clip, click and drag the green slider (for the start frame) to the desired location—MyDVD updates the starting thumbnail preview image so that you can tell where you are within the video clip.

5. Select the ending point for your clip by clicking and dragging the red slider (for the end frame) to the desired location—again, the ending thumbnail preview image is updated as a reference.

FIGURE 6.9   Trimming excess from the beginning and ending of our clip.

tip **For more precise control, click on the slider you want to move and press the left and right arrow keys to move backward and forward.**

6. Satisfied with the trimmed video clip? If so, click OK to return to the menu. To cancel your changes without saving them, click Cancel— or, to start over with the trimming process, click Reset.

7. If you like, you can specify the thumbnail you want to display from anywhere within the video clip.

To delete a movie button, right-click on it and choose Delete Movie—note that this simply removes the button and the linked video clip, it does not delete your original video clip from your hard drive.

tip **MyDVD has a limit of six buttons (of either type) on a single menu; if you attempt to add more than six buttons, the program automatically displays a dialog explaining that a new *continue* menu has been created, and adds your button on the new menu. (In this case, *continue* means that the new menu is not a submenu but a continuation of the previous menu.) You can always tell when you're on a continue menu, because MyDVD adds the previous and next buttons to the lower right corner of the menu. You can double-click on these navigational buttons to move to the previous and next menus in the continuous series.**

Changing Menu Styles

In MyDVD, a *menu style* is a combination of these separate choices:

- The background image
- The border around menu and movie buttons
- The fonts
- The color scheme

Although Allegro, the default menu style, is a great choice for a DVD with video clips from a piano recital, it's certainly not appropriate for a birthday party! Luckily, MyDVD doesn't restrict you to just one menu style; there are 27 preset menu styles to choose from, with designs for holidays, family events, and themes such as TV shows, nature trails, and rainy days. You can also create your own custom menu styles by combining elements from the preset menu styles.

✔ **To change the menu style for your MyDVD project at any time, using one of the preset styles, follow these steps:**

1. Click the Style button at the top right corner of the menu editor window. (You can also press Ctrl+L.) MyDVD displays the Select a Style dialog, as shown in Figure 6.10.

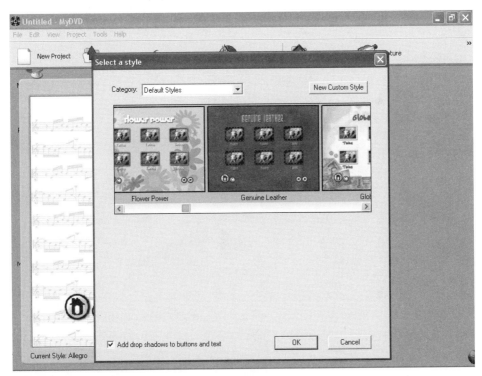

FIGURE 6.10   Choosing a preset menu style.

2. Scroll through the style thumbnails until you find the one you want to use with this project. Click on it to highlight it.

3. You can specify whether you want to add a 3-D drop shadow effect to the buttons and text on your menus; by default, the option is turned on. (Personally, I prefer it with the drop shadows, but it makes the text hard to read on some televisions.) To turn this option off, click the Add drop shadows to buttons and text checkbox to disable it.

4. Click OK to use the new menu style; Figure 6.11 illustrates the "Halloween 01" menu style.

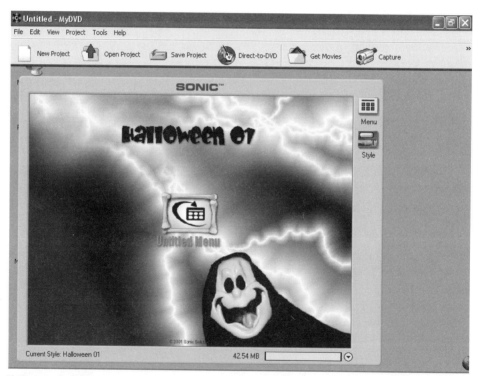

FIGURE 6.11 With a click of the mouse, our project is ready for Halloween!

✔ **Follow these steps to create a custom menu style:**

1. Click the Style button at the top right corner of the menu editor window to display the Select a Style dialog.

2. Click the New Custom Style button to display the New dialog, illustrated in Figure 6.12, and type a new name for your custom style. Click OK to continue.

3. MyDVD displays the dialog you see in Figure 6.13—note that the Category has changed to Custom. Click the Browse button to select a background, and navigate to the desired Windows bitmap or JPEG image. Click it to select it and click Open. MyDVD updates the thumbnail preview image to reflect your change.

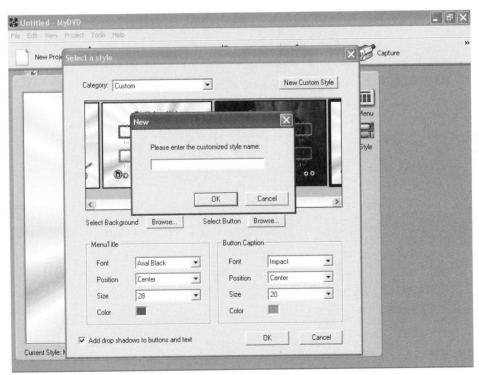

FIGURE 6.12 Entering a new name for a custom menu style.

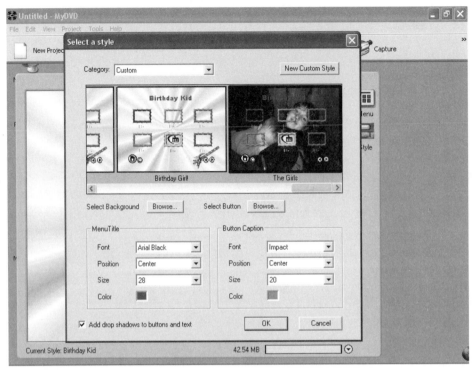

FIGURE 6.13   Preparing a custom menu style.

4. Click the Browse button to select a button border style. MyDVD displays the selection of buttons you see in Figure 6.14; click the appropriate button style and click OK. MyDVD updates the thumbnail preview image to reflect your change.

5. Click the drop-down list boxes in the Menu Title group to specify a font, position, and size for the menu titles. To choose the color of the menu font, click the color square to display the standard Windows color palette selector shown in Figure 6.15; click the color you want, then click OK.

tip **Editing the menu title text is as easy as changing the captions on your menu buttons: Simply click on the title and type the new text.**

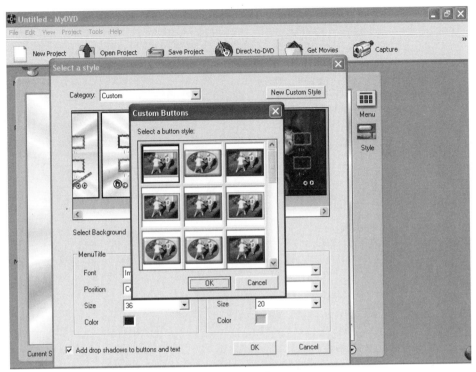

FIGURE 6.14 Choosing a button border style.

6. Click the drop-down list boxes in the Button Caption group to specify a font, position, and size for the captions that appear under your
menu buttons. To choose the color of the caption font, click the
color square to display the (now familiar) standard Windows color
palette selector; click the color you want, then click OK.

7. By default, MyDVD adds a 3-D drop shadow effect to the buttons
and text on your menus—to turn this option off, click the Add drop
shadows to buttons and text checkbox to disable it.

8. Click OK to save the custom menu style; Figure 6.16 illustrates a
custom style I created for a video tour of WWII military airplanes,
using my own photo as a menu background. You can now choose
the custom style any time you display the Select a Style dialog.

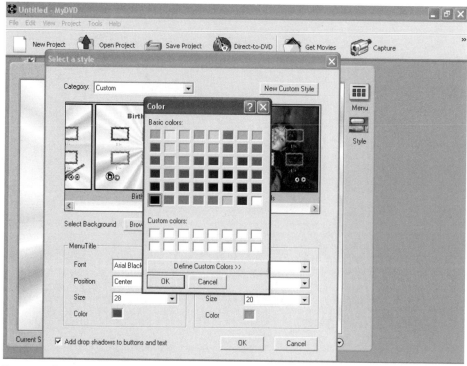

FIGURE 6.15   What color goes best? Use the Windows color selector to choose your color.

FIGURE 6.16 My custom menu background.

Checking the Safe Zone

So what's a *safe zone*, anyway? That's a term that video editors use when referring to the dimensions of your television screen. You may not know it, but the image designated by the signal that's beamed to your television is actually larger than the television screen; the additional area around the outside of the image that you can't see takes care of distortion (call it the "ragged edge" of the TV image).

When you're creating a MyDVD disc, the safe zone refers to that part of the video clip that will actually appear on screen, and it's important to verify that your movies are correctly displayed in the safe zone before you record them. MyDVD can display a rectangle that signifies the safe zone: To display it, click View and choose the Show TV Safe Zone menu item. Figure 6.17 illustrates the dashed line that appears to indicate the safe zone.

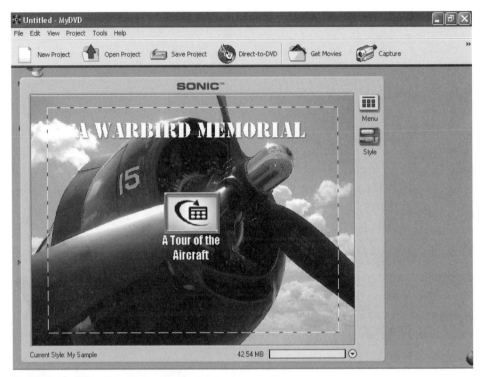

FIGURE 6.17   Displaying the safe zone around a menu.

To hide the safe zone border, click View and choose the Show TV Safe Zone menu item again.

tip **As a general rule, while you're filming, try to keep the subjects of your movies in the center of the frame—this will keep them from being "chopped" outside the edge of the safe zone.**

Previewing Your Work

Okay, you've created all of your menus, added and trimmed all of your movies, edited the text of your menu titles and buttons, and checked the border of your project, using the safe zone display. Are you ready to burn? Technically, you are indeed finished and ready to record your disc; however, MyDVD offers a Preview player that can display your project as it will appear on the finished disc, complete with a "virtual" remote control that simulates the remote for your DVD player!

To enter Preview mode at any time, click Preview on the MyDVD toolbar (or press Ctrl+P). The project appears in the menu editor, with the remote control pad shown in Figure 6.18. Click on the remote control buttons just as you would press the buttons on your DVD remote control.

To jump directly to the Title menu, click the Title button—clicking the Menu button displays the last menu you used.

When you're done previewing your project (and you want either to record the disc or to edit the project further), click the Close button on the remote control.

FIGURE 6.18 Doing a little testing in Preview mode before I record.

Recording the Disc

When you're ready to record your project, MyDVD gives you three choices:

1. **Make a CD.** If you record your project to a CD-R or CD-RW disc, MyDVD creates what's called a *cDVD* (a trademark of Sonic, the makers of MyDVD)—these discs will play on virtually all computer CD-ROM and DVD-ROM drives, using a free viewer program that MyDVD automatically adds to the disc.

2. **Make a DVD.** If you record using a DVD+RW or DVD+R disc, your disc will play in most DVD players and computer DVD-ROM drives.

3. **Make a DVD to hard disk.** Select this option to create a *DVD volume* in the folder you specify, which is something like a disc image—although a DVD volume can be recorded to a disc later, it's often used to create a DVD volume for use with software DVD players, such as PowerDVD and WinDVD.

✔ Follow these steps to record your project:

1. Load a blank disc of the proper type into your recorder—no disc is necessary if you're going to create a DVD volume.

2. Click the Make Disc button on the MyDVD toolbar, and MyDVD prompts you to save the project. Choose a filename and a location, and click Save.

3. The Make Disc Setup dialog shown in Figure 6.19 appears. Click the desired media type; if you have more than one recorder on your system, you can also choose which drive will be used. Leave the Write Speed setting at Auto.

4. Click OK to begin recording!

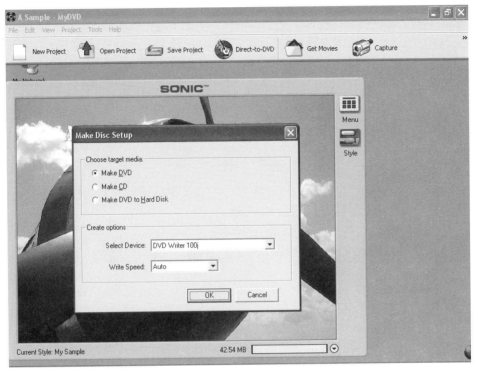

FIGURE 6.19 Preparing to record my project.

Project: **Recording a DVD Video Disc with MyDVD**

In this project, we'll take video clips that you've stored on your hard drive and create a family DVD Video disc that you can watch on your DVD player.

Requirements

- MPEG, AVI, or MOV digital video clips
- Installed copy of MyDVD
- A blank DVD+RW or DVD+R disc

✔ **Follow these steps:**

1. Click Start | All Programs | MyDVD | MyDVD to run the program.

2. From the initial Wizard menu, click the New DVD Project option and click Next to continue.

3. The Wizard prompts you to choose a menu style for your project (Figure 6.20). For this project, I'll click the Summer Fun thumbnail to select it. Click Finish, and the MyDVD toolbar and menu editor appear with the style you've selected.

FIGURE 6.20 Selecting a menu style using the Wizard.

4. For this project, I'll add two buttons: One will feature a clip with the family dog, and the other will have two clips from summer vacation. Click twice on the Menu button to add two buttons to the layout.

5. Let's change the menu title and the captions for both buttons. Click on the menu title, type "Our Family's Summer 2002," and press Enter to save the change. Click on the caption for the left button and type "Let's Wash the Dog"—press Enter to save the change. Finally, click on the right button caption and type "Summer at the Beach" and press Enter. Figure 6.21 illustrates our completed Title menu.

6. Next, double-click the left button you added—note that MyDVD automatically updates the title of this menu to match the button that links to it. Click the Get Movies button on the MyDVD toolbar. Navigate to the location of your video clip, click it once to highlight it, and click Open to add the video button.

FIGURE 6.21 The Title menu looks great.

7. The clip you've just added has a 4-second title and transition that we no longer need, so double-click on the movie button to display the Trimming dialog you see in Figure 6.22. Note that I've moved the green slider 4 seconds to the right to eliminate the beginning title. Click OK to continue.

8. To return to the Title menu, double-click on the Home button.

9. Double-click the right button you added. We'll add two movies to this menu. Click the Get Movies button on the MyDVD toolbar. Navigate to the location of the clips and hold down Ctrl while you click both filenames, then click Open to add two video buttons. Double-click on the Home button to return to the Title menu.

10. Next, let's check the safe zone for our project: Click View and click Show TV Safe Zone to toggle this feature on. As you can see in Figure 6.23, our title and buttons fit well within the safe zone border.

FIGURE 6.22 Trimming the first 4 seconds from one of our video clips.

FIGURE 6.23 Checking the safe zone around our menu.

11. Finally, let's preview our disc before we record it—click Preview on the MyDVD toolbar, then use the controls to change menus and run each of the movies. (Note that the remote control also displays the elapsed time for the clip you're viewing.) If everything checks out okay, click the Close button on the MyDVD remote control to return to the menu editor.

12. We're ready to record! For this project, I'll load a blank DVD+RW disc into my drive.

13. Click the Make Disc button on the MyDVD toolbar, and MyDVD prompts you to save the project. Choose a filename and a location, and click Save.

14. MyDVD displays the Make Disc Setup dialog. Click the DVD option.

15. Click OK and sit back while your DVD masterpiece is created.

Project: **Recording a cDVD with a Custom Menu**

As I mentioned earlier, MyDVD can create a cDVD using a CD-R or CD-RW disc; I'll design and build a cDVD in this project, using a custom menu design.

Requirements

- MPEG, AVI, or MOV digital video clips
- Installed copy of MyDVD
- A blank CD-R or CD-RW disc

✔ **Follow these steps:**

1. Click Start | All Programs | MyDVD | MyDVD to run the program.

2. From the initial Wizard menu, click the New DVD Project option and click Next to continue.

3. The Wizard prompts you to choose a menu style for your project—because we're going to change the design, you can select any style (I used the first one, Allegro), then click Finish.

4. Click the Style button at the top right corner of the menu editor window to display the Select a Style dialog.

5. Click the New Custom Style button to display the New dialog, and type a new name for your custom style. Click OK to continue, and click the new style thumbnail to select it.

6. On the Select a Style dialog, click the Browse button to select a background, and navigate to the desired Windows bitmap or JPEG image (Figure 6.24). Click it to select it and click Open; in this case, I'll choose one of the default background bitmap images that comes with Windows XP.

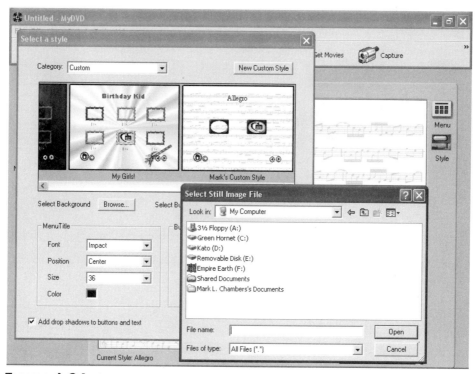

FIGURE 6.24 Locating a background image for our custom style.

7. Next, click Browse and select a button border style. I'll click the light blue oval button border. MyDVD automatically updates the thumbnail preview image.

8. For this project, I'll choose the Rockwell font for my menu text, and I'll set the font size at 28.

9. Because I like matching fonts on a menu, let's also click the Font drop-down list box for the button captions and specify Rockwell; however, I want white button captions, so click the color square to display the standard Windows color palette selector. Click white and click OK to save the change.

10. We're done! Click OK to save the custom menu style; the new design is shown in Figure 6.25 (perfect for a camping trip video).

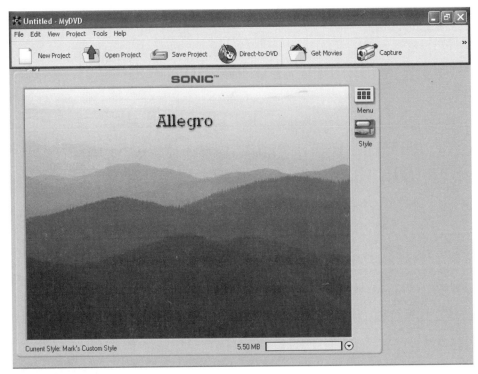

FIGURE 6.25   Our new custom style.

11.   I have only one clip to add to this disc (because it's a cDVD disc, we have less space to work with)—it will feature digital video that I took on our last family camping trip. Click on the Menu button to add a single button to the layout.

12.   Next, it's time to change the menu title and the captions. Click on the menu title, type "Roughing It" and press Enter to save the change. Click on the caption for the button and type "Camping in the Ozarks"—press Enter to save the change. Figure 6.26 illustrates our completed Title menu.

13.   Double-click the button you added and click the Get Movies button on the MyDVD toolbar. Navigate to the location of your video clip, click it once to highlight it, and click Open to add the video button.

14.   That completes our menu design! To return to the Title menu, double-click on the Home button.

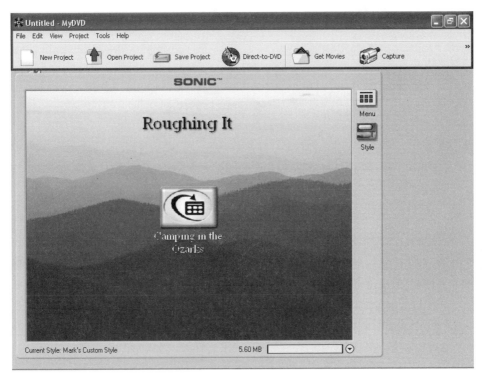

FIGURE 6.26 The camping video Title menu.

15. Click View and click Show TV Safe Zone to make sure that both the title and the button fit within the safe zone border—no problem with this project, because we needed to add only a single button.

16. We're ready to burn—but first, click Preview on the MyDVD toolbar and use the remote control to check the menu and watch the video. If you like what you see, click the Close button on the MyDVD remote control to return to the menu editor.

17. Because we're recording a cDVD, load a blank CD-R disc into the drive.

18. Click the Make Disc button on the MyDVD toolbar, and MyDVD prompts you to save the project. Choose a filename and a location, and click Save.

19. MyDVD displays the Make Disc Setup dialog. Click the cDVD option.

20. Click OK, and recording begins!

Summary

In this chapter, you learned how to design, edit, and produce your own DVD and cDVD discs using MyDVD and digital video clips stored on your hard drive.

7

Direct-to-DVD Recording

In This Chapter

✔ Connecting your equipment

✔ Selecting capture settings

✔ Setting chapter points

✔ Project: Recording a DVD from a DV Camcorder

In the previous chapter, you learned how to record a DVD (or cDVD) disc using MyDVD and digital video files from your hard drive—this is likely to be the route that most of us will take with our digital video footage.

However, MyDVD also offers a direct-to-DVD mode, where the video input from sources such as a digital camcorder or VHS VCR can be burned directly to a DVD. No editing required!

In this chapter, I'll discuss the process of direct-to-DVD recording using MyDVD, and we'll burn a DVD disc using this technique (with the help of the MyDVD Wizard). By the way, you can also record a CD-R or CD-RW directly from your DV source, but I wouldn't recommend it unless you're going to be recording less than 30 minutes of uncompressed footage.

Connecting to the Source

In videospeak, a *source* (or, as MyDVD calls it, a *video device*) is any piece of hardware that can send one of two types of signals to your computer:

1.   A digital video signal, using a FireWire (or IEEE-1394) connection to your computer. This is the method you'll use to connect a DV camera to your PC. This type of connection is pure joy: It's *plug-and-play* (meaning that your PC automatically detects that you've connected your DV camcorder), and you don't have to reboot after you plug in the FireWire cable.

2.   An analog video signal, using a DirectShow-compatible video capture device. You'll need this hardware if you want to connect a VCR, VHS camcorder, or TV (with Video Out) to your PC.

tip **Do you have an older PC that doesn't have a built-in FireWire connector? Don't panic—no need to buy another computer just to connect your DV camcorder! Instead, you can add a FireWire card to your existing PC; these cards usually cost under $100. You'll also need an open PCI slot in your computer where you can install the card. (Most video capture cards will also need the same type of PCI slot.) If your recorder uses a USB 2.0 connection, you can also install a FireWire/USB 2.0 combo card and take care of both requirements with one piece of hardware.**

Unfortunately, there's no one "standard" way to connect an analog device—you'll have to refer to the manual that accompanied your video capture card to determine how to connect your analog video source to your computer. Note also that you may require a separate connection for the analog audio signal, as well.

Configuring Windows for Direct-to-DVD Recording

Unlike the recording from files that I covered in the last chapter, direct-to-DVD recording is much more demanding on your entire system—therefore, it's important to make sure that your computer and its operating system have been correctly optimized for best performance. In this section, I'll discuss the three important steps you should take before attempting a direct-to-DVD session.

Using DMA Mode

DMA stands for *Direct Memory Access*, and it's the fastest method of transferring data directly from your computer's memory to your recorder (without processing it or routing it through the CPU beforehand). To record directly from a video source to your recorder successfully, you should enable DMA on your primary hard drive.

✔ Follow these steps:

1. Log onto Windows as Administrator (or as a user with Administrator privileges).

2. Right-click on My Computer and choose Properties from the pop-up menu to display the System Properties dialog (Figure 7.1).

3. Click the Hardware tab to display the hardware controls (Figure 7.2).

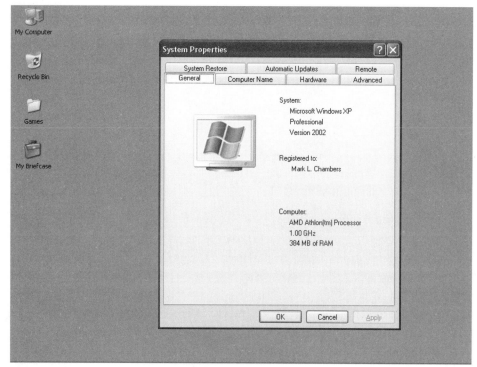

FIGURE 7.1   The System Properties dialog within Windows XP Professional.

4. Click Device Manager to display the window you see in Figure 7.3.

5. Click the plus sign next to the IDE ATA/ATAPI controllers entry to expand it, as shown in Figure 7.4. (Note that the hardware listed under the entry may not exactly match what you have on your system.)

6. Right-click on the Primary IDE Channel entry and choose Properties.

7. Click the Advanced Settings tab to display the options you see in Figure 7.5.

8. Click on the Transfer Mode drop-down list for Device 0 and choose DMA if available.

9. Click OK to return to the Device Manager, then click the Close button on the Device Manager window to return to Windows.

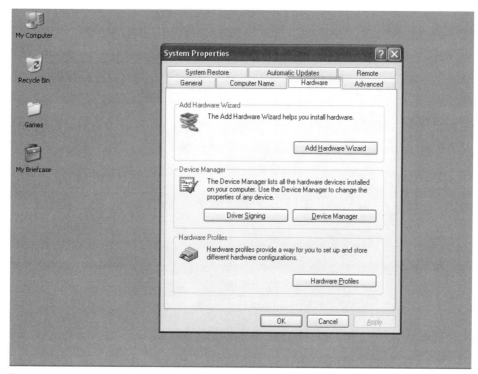

FIGURE 7.2   The System Properties Hardware panel.

Defragmenting Your Drive

I discussed how and why you should defragment your hard drive in Chapter 3, so I won't go into it here; suffice it to say that you'll need to fine-tune and optimize your hard drive to get the top performance you need for error-free direct recording.

If you missed my sage advice on defragmenting—as well as the step-by-step procedure to follow—turn back to the section titled "Defragmenting Your Hard Drive" in Chapter 3. We'll wait for you here!

Avoiding Multitasking

Finally, it's very important to avoid running any other programs at the same time you're using MyDVD to record direct-to-DVD. Multitasking

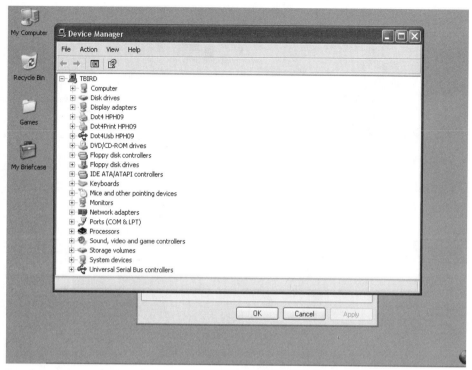

FIGURE 7.3   The Windows XP Device Manager revealed.

with a "performance sloth" such as Word or Photoshop, for example, will be a serious drain on your PC's memory and hard drive resources; the older the computer, the more likely that this performance hit will adversely affect the recording session and MyDVD will abort the recording.

Therefore, before you begin direct-to-DVD recording, make sure that MyDVD is the only application running under Windows, and shut down as many of the background programs running in the Windows taskbar as possible.

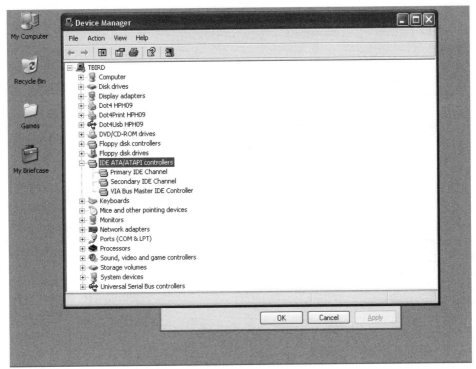

FIGURE 7.4   Displaying the controllers on your system.

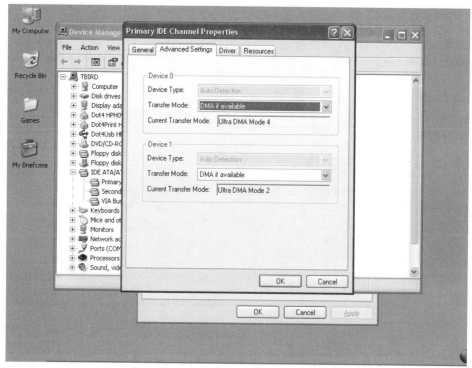

FIGURE 7.5   Enabling DMA transfer mode for my hard drive.

Selecting Capture Settings

If you've connected a single FireWire device, MyDVD will automatically recognize it as a source, and you can skip this section—it's possible, however, that you have either connected an analog device or you have multiple sources connected to your PC. If so, you must select the video and audio devices that Windows will use as the source for your recording. Note that you must have already successfully installed the capture device within Windows, or the connection won't show up during this procedure.

tip It's very important to keep your capture hardware updated with the latest Windows drivers—check the capture hardware manufacturer's Web site at least once a month. Current drivers will go a long way toward assuring trouble-free operation—that applies to all of your PC's components!

✔ **Follow these steps:**

1. Click Tools and choose the Capture menu.

2. Click Change to display the current video device—to select a new device, click on the entry that corresponds to the hardware you've connected.

3. If the video device you've chosen has multiple input connectors—for example, a capture card that can handle multiple sources—choose the entry for the input you'll use from the source list.

4. If the Configure button is enabled, click it and set the video configuration to NTSC, 720 × 480, and use either I420 or IYUV for the color model/image format setting. Click OK to save your changes. (Note that these settings apply within North America, whereas many other countries around the world use PAL or SECAM, a derivative of PAL. The software defaults to NTSC; however, if you change the selection, it will default to the last setting used. MyDVD will inform you if you are trying to import a PAL clip into an NTSC project; you can convert clips between NTSC and PAL using ShowBiz.)

5. If the Select button is enabled, click it and set the configuration to the input you're using and NTSC. Click OK.

6. Click on the entry that corresponds to the audio device you'll be using.

7. If the audio device you've chosen has multiple input connectors, choose the entry for the input you'll use from the source list.

8. Click OK to save your changes and return to MyDVD.

Whew! You can see why the DV-FireWire route is so popular in the world of video recording; because it's automatic, you can avoid a lot of these settings altogether.

Setting Chapter Points

MyDVD allows you to set chapter points for your disc that operate just like those in a commercial DVD movie: You can jump directly to a chapter point at any time. MyDVD can create chapter points automatically at timed intervals in your recording—just enable the Create Chapter Points checkbox within the direct-to-DVD Wizard, and choose the interval time (either in seconds or minutes). When capturing from DV, the program can be set to create chapter points automatically where the tape was started and stopped.

 tip   **To set a chapter point manually while you're recording, press the spacebar.**

Project: **Recording Directly from a DV Camcorder**

Let's record a DVD directly from a DV camcorder, using MyDVD and a FireWire connection.

Requirements

- DV camcorder and FireWire cable
- Installed copy of MyDVD
- A blank DVD+RW (or DVD+R disc, if you're using the DVD-Writer DVD200)

✔ **Follow these steps:**

1. Click Start | All Programs | MyDVD | MyDVD to run the program.

2. From the initial Wizard menu, click the Record Direct-to-DVD option and click Next to continue.

3. The Wizard prompts you to enter a name for the project—type a name and click Next to continue.

4. Next, the Wizard displays the screen you see in Figure 7.6. Because we're recording to a DVD disc, click Make DVD. Click on the Select Device drop-down list and choose your drive; leave the Write Speed option set to Auto. Click Next to continue.

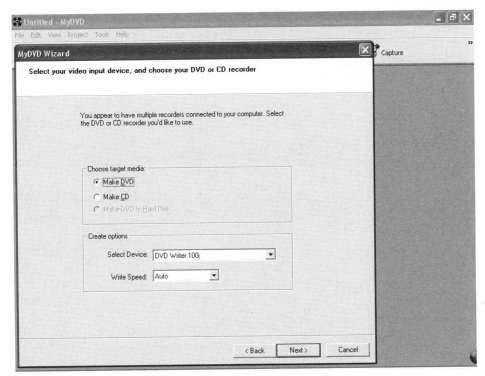

FIGURE 7.6 Choosing the target drive and media.

5. Choose a menu style by clicking on the thumbnail—or, if you prefer to burn a disc without menus, click No menus. You can add drop shadows to buttons and text if you like. Click Next to continue.

6. The next screen allows you to record just the video signal from the source—a good option when audio isn't important, because it will save a significant amount of space on the disc—as will the video quality option you choose. (At Best quality, you can fit only about 60 minutes of video on a 4.7-GB DVD; at Good quality, you can fit a whopping 180 minutes.) You can also set MyDVD to create chapter points automatically.

7. Because we're using a FireWire connection as our source, you can actually control your DV camcorder through the Wizard by using the VCR-style controls; note that you can move backward one frame by clicking the minus button, and you can move forward one frame by clicking the plus button. Using these controls, move to the frame where you want to begin the recording.

8. Click Record—MyDVD will automatically start the camcorder.

9. MyDVD will buffer the signal until the disc limit has been reached, then the actual recording process will begin.

Summary

This chapter covered direct-to-DVD recording using MyDVD; I discussed the requirements for both digital and analog video recording, along with the steps you should take to optimize Windows before recording. Finally, we recorded a disc using a DV camcorder as a source.

8

Watching DVD Video with PowerDVD

In This Chapter

✔ Loading and playing a DVD movie

✔ Using the standard DVD player controls

✔ Using bookmarks

✔ Viewing subtitles

✔ Watching DV files from your hard drive

✔ Project: Capturing and E-mailing Images from Digital Video

When you bought your DVD recorder, you may not have considered the primary "fringe benefit" of that purchase: the ability to watch DVD-Video discs on your computer. In fact, many videophiles prefer the sharper picture you'll get with a computer monitor over a standard TV set; if you've spent the money on a sophisticated surround sound speaker system and subwoofer for your PC, you may also enjoy better audio during the movie by watching it on your computer. These days, some DVD movie discs even come with programs such as trivia games and 3-D models that you can run only if you've loaded the disc in a DVD drive.

I n this chapter, I'll use the PowerDVD software viewing program that accompanies the HP DVD-Writer drive: You'll learn how to operate the same "virtual" controls on your screen (using your mouse and your keyboard) that you would find on an actual DVD player remote control. In addition, I'll also demonstrate how you can display digital video files straight from your computer's hard drive—something that your mundane DVD player can't do—and how you can capture images directly from the display and e-mail them to friends and family. Gather the family around that 19-inch monitor and fix a bowl of popcorn—PowerDVD is about to turn your computer into a movie theater!

Loading and Playing a DVD Movie

Your set-top DVD player is simple to operate, and PowerDVD is just as easy to run. There are three methods of starting PowerDVD:

1. **From the Start menu.** Click Start | All Programs | Cyberlink PowerDVD | PowerDVD.

2. **From the desktop shortcut.** The PowerDVD installation program places a shortcut icon on your Windows desktop—double-click it to start the application.

3. **Automatically upon loading a DVD.** You can configure Windows XP to run PowerDVD automatically each time you load a DVD movie. The first time you load a movie, you'll see the infamous "What Should I Do?" dialog (shown in Figure 8.1). Click the Play DVD Movie using PowerDVD entry from the list and click the Always do the selected action checkbox to enable it, then click OK.

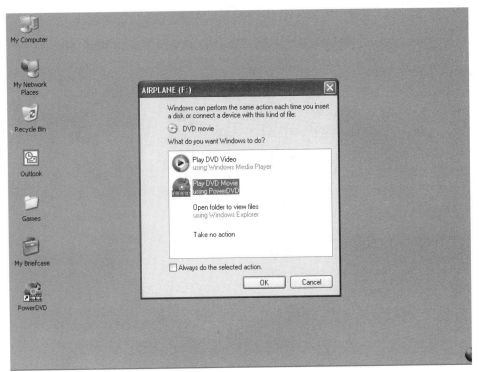

FIGURE 8.1   Configuring Windows XP to run PowerDVD automatically when you load a DVD movie.

Once PowerDVD is running, you'll either see the program's display window and control panel (shown in Figure 8.2) or the movie will begin playing automatically—the result depends on the setting you've selected in the General configuration dialog, which I'll explain later in the chapter). If the movie doesn't begin playing automatically, check the Mode button and make sure PowerDVD is in Disc mode (click the Disc/File Mode button to cycle to Disc mode, if necessary), then click the Play button; Figure 8.2 also shows both of these controls. When the DVD movie's menu appears, click on the desired menu option to select it.

If all you want to do is simply watch a movie, that's all there is to it! However, if I stop here, we're left with a two-page chapter, and you won't learn about all of the other great features included in Power-DVD. Therefore, watch a movie or two, or three (if you have to get

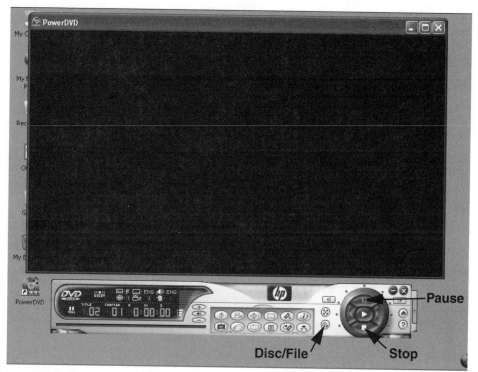

FIGURE 8.2   PowerDVD is ready to go.

the novelty out of your system first), then come back, and I'll describe the rest of the program.

Basic Video Controls Explained

Most of the controls you'd expect to find on a good DVD player are available within PowerDVD, along with several features that are unique to the PC that may be unfamiliar to you. In this section, I'll cover these common controls and demonstrate how they're used.

tip **PowerDVD provides a control menu that's always active—even in full-screen mode, when you normally can't see the control panel. To display the control menu, right-click the mouse and choose the desired option.**

Note that PowerDVD usually provides both *on-screen* controls (on either the control panel or the right-click menu) and *keyboard* controls (which you can use at any time) for each feature.

Showing and Minimizing the Panel

Although the control panel is handy to have around, it can get in the way of the action when you're watching a movie full-screen. To minimize the panel, click the button with the minus sign at the top right corner of the control panel, as shown in Figure 8.3 (or press the Ctrl+N keyboard shortcut).

 tip **If you're viewing a movie full-screen and the Ctrl+N shortcut doesn't work, use Alt+Tab to switch to the panel.**

Switching Between Window and Full Screen

Personally, I prefer watching my movies full-screen, which takes advantage of my 19-inch monitor; however, if you like to leave a movie running in a window, it's possible to work on a document or use another application while you're screening an old favorite from your DVD collection. (That is, of course, if you can actually concentrate on your work while Sigourney Weaver is toasting aliens with a flamethrower.)

Like any well-mannered Windows program, you can resize the PowerDVD display window by dragging the right corner of the window border; PowerDVD will automatically adjust the resolution and aspect ratio of the display window to make the best use of the new window dimensions.

To switch PowerDVD between full-screen and windowed display, use the Zoom button (Figure 8.3), or display the right-click menu and choose Zoom.

FIGURE 8.3   The PowerDVD window controls.

Pausing and Stopping the Movie

It never fails: The phone rings, there's a meeting to attend, or you have to sign for a FedEx package just as the action really gets good. You don't have to miss a second of your movie if you pause PowerDVD: If the control panel is visible, click on the Pause button (Figure 8.2), or display the right-click menu and choose Pause. You can also press the spacebar to pause the movie, or click on the Shuttle dial at the top position (more on the Shuttle dial in the next section).

If you need to stop the movie completely, you can either:

- **Exit PowerDVD.** Press Ctrl+X or click the Exit button on the control panel, as shown in Figure 8.3. (You can eject the disc by pressing Ctrl+E.)

- **Stop the movie.** You can also stop the movie without exiting PowerDVD. Click the Stop button (Figure 8.2), press S or right-click and choose Stop.

While the movie is paused, you can step forward or backward one frame at a time—this is a great way to view a particularly complex or extremely fast action sequence in "slow motion." (For example, I'll step forward quite a bit while watching one of my Bruce Lee movies!) To step, click either the Step Backward or Step Forward buttons (shown in Figure 8.4), or display the right-click menu and choose Forward or Backward. The key sequences are Ctrl+B for a backward step and T for a forward step.

Step Backward   Step Forward

FIGURE 8.4   Use these PowerDVD controls to navigate through a movie.

Fast-Forward and Rewind

Most of today's DVD players offer a circular control called a *Shuttle dial* that you use to fast-forward and rewind through the movie while it's playing, and PowerDVD uses a Shuttle dial, as well (as shown in Figure 8.4). The Shuttle dial is active only when the movie is playing.

To use the dial, just click on one of the dots that appear around the outside edge: Each dot corresponds to a specific speed in either the forward or reverse directions. For example, clicking the outside edge of the dial at the top position—the one directly above the Pause button— actually has the same effect as the Pause button. Clicking any position to the left of this top position rewinds the movie, and the farther down along the edge of the dial you click, the faster the rewind speed. Clicking the right side of the dial moves forward—again, the farther down the edge of the dial, the faster the forward speed.

To help you visualize the operation of the Shuttle dial, Figure 8.4 illustrates four positions: 8X rewind at the bottom left of the dial, 1X rewind at the top left, 1X forward at the top right, and 8X forward at the lower right. The other dots along the edge of the dial correspond to 2X and 4X rewind and forward speeds.

When you click on the Shuttle dial, PowerDVD places a green dot at the position to indicate the shuttle direction and speed.

Selecting a Chapter

Today's DVD movies are divided into *titles* and *chapters*, making it easy to start at specific points in the movie or jump from one part of the film to another to view your favorite scenes. A title is a named section of the film—for example, in *Gone With the Wind*, one title might read "A Party at Tara"—and each title usually includes a number of chapters, each of which you can jump to directly. Most of the DVD movies available today use an interactive menu that appears before the movie actually begins, and one menu item allows you to watch the movie beginning at any chapter point you choose.

Because the DVD interactive menu appears when I load a movie, I usually use it to make my initial chapter choice; PowerDVD displays the menu, and you simply click on the desired chapter thumbnail image. However, there's a faster alternative you can use to select a specific chapter: Use the PowerDVD shortcut menu. Right-click on the Play button at the center of the Shuttle dial, and PowerDVD displays the chapters in the film (organized as submenu items under the title names, as shown in Figure 8.5).

Unfortunately, the shortcut menu doesn't display the thumbnail images you'll get from the DVD menu's built-in chapter selector…but if you know a film by heart, you may even learn the chapter numbers, as well!

You'll note that the shortcut menu also displays a Browser option— select it, and PowerDVD opens the dialog you see in Figure 8.6. The Browser is essentially a graphical representation of the same title and chapter hierarchy that's available from the shortcut menu, but I find it easier to use; to choose a specific chapter, just double-click on it. (You can also use your cursor keys to navigate the Browser tree.)

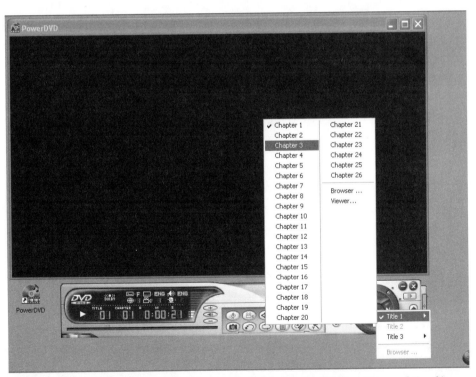

FIGURE 8.5   The titles and chapters available for the DVD edition of the film *Airplane!*

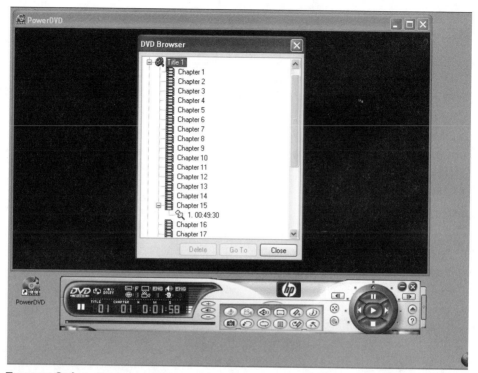

FIGURE 8.6   Selecting a chapter from the Browser dialog.

Using Next and Previous

Are you tired of a particular scene but don't want to jump elsewhere in the film or lose the flow of the plot? You can skip from one chapter to the previous or next chapter using the Next and Previous buttons (shown in Figure 8.4) or display the right-click menu and choose Next or Previous from there. The keyboard shortcut buttons for Next and Previous are N and P, respectively.

Using Repeat

You can repeat the display of an individual chapter or a title: Click the Repeat button (Figure 8.7) once to repeat (or loop) the current chapter. Click the button twice to repeat the current title; clicking the Repeat button again turns the function off. You can press Ctrl+R from the keyboard, as well.

PowerDVD doesn't stop there, however; you can set your own specific scenes to repeat, using the program's AB Repeat feature.

✔ **Follow these steps to set a custom repeat while you're watching a movie:**

1. Click the AB Repeat button (Figure 8.7) or press X to mark the start of the segment (point A).

2. When the movie reaches the point where you want the segment to end, click the AB Repeat button or press X again to set the end of the segment (point B). PowerDVD will immediately begin looping the segment you've marked.

3. To cancel the AB repeat, click the AB Repeat button again.

FIGURE 8.7   The PowerDVD repeat and audio controls.

Setting the Volume Level

Although you can set the volume level in Windows using the taskbar Volume Control—display the taskbar at the bottom of the screen and click on the speaker icon—PowerDVD offers a more elegant and convenient solution with its own volume controls. Figure 8.7 illustrates the three audio level controls:

1. Click the Increase Volume button or press the plus (+) key on your keyboard to turn up the audio.

2. Click the Decrease Volume button or press the minus (-) key on your keyboard to turn down the volume.

3. Click the Mute button or press the Q key on your keyboard to temporarily turn off the audio altogether.

Selecting an Audio Stream

Many DVD movies feature a number of different *audio streams*—for example, a movie can include the dialog spoken in French or Spanish, commentary from the actors and director, or even a complete alternate soundtrack. Some discs may also include specialized audio streams with playback modes such as Dolby surround sound or Dolby Pro-Logic (whereas older discs offer only plain-and-simple stereo).

The selection of audio streams that you can choose from in PowerDVD while watching a DVD movie depends on three criteria:

1. The sound card you've installed in your computer

2. The speaker layout you're using

3. The DVD movie itself (what audio streams the studio decided to add for this specific movie)

You can choose an audio stream from the DVD's own interactive menu, or you can put PowerDVD to work. Click the Audio Streams button shown in Figure 8.7 to toggle between the different streams, or press H.

Switching Angles

If you're watching a DVD disc that features multiple camera angles, click the Angle button shown in Figure 8.8 to switch between the different angles, or press A.

Bookmarking Scenes

Earlier, I mentioned AB Repeat; in that same vein, *bookmarking* a scene is another PowerDVD feature designed to make it easy to locate a specific scene in your favorite movie. To add a bookmark, you click the Set Bookmark button on the control panel (Figure 8.8) or press Ctrl+F2 at the desired moment in the film; you can do this as often as you like, and PowerDVD keeps track of each bookmark.

FIGURE 8.8   PowerDVD's bookmark, angle, and subtitling controls.

There are four methods of selecting a bookmark that you've set:

1. Press the F2 button or click the Go To Bookmark button to toggle through your bookmarked scenes.

2. Right-click the Go To Bookmark button and choose a bookmark from the pop-up menu.

3. Right-click the Go To Bookmark button and choose Browse to display the Browse dialog, which looks very much like the Chapter Browse dialog—double-click on the desired bookmark (they look like pushpins, and they appear underneath the chapters).

4. Right-click the Go To Bookmark button and choose Viewer (for the thumbnail viewer you see in Figure 8.9)—double-click on the desired bookmark thumbnail.

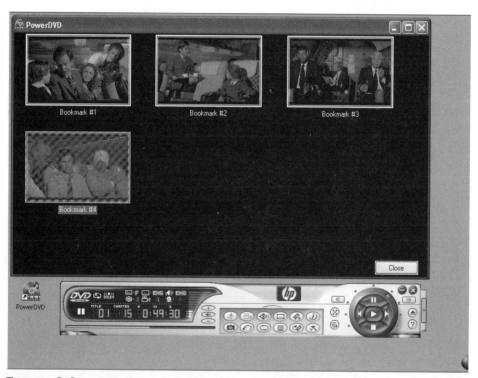

FIGURE 8.9   *Selecting a bookmarked scene using the Viewer.*

Displaying Subtitling

Like audio streams, most of today's DVD movies feature subtitles—usually, you'll get subtitles in English and one or two other languages, but I've seen as many as five different subtitle selections for one film. Figure 8.8 illustrates the Subtitles button, which you can click while the movie is playing to toggle between the different subtitles available on the disc—from the keyboard, press the U key to toggle the subtitles.

You can also right-click on the Subtitles button to display a pop-up menu—click on the specific subtitling you want from the menu to select it.

tip **If your DVD movie also provides closed captioning, you can toggle this feature on, using the Subtitles button; right-click on it and choose Closed Captioning from the Secondary Subtitles section.**

Using File Mode

As I mentioned at the beginning of the chapter, PowerDVD isn't limited to displaying only DVD movies; you can also listen to audio CDs and display video CDs (also called *VCDs*) using the same controls as DVD movies—a neat feature and one you'll likely use often.

But why limit yourself to discs? You can view MPEG, QuickTime MOV, and Windows AVI-format video directly from your hard drive! (In fact, most DVD movies are "assembled" and burned from hard drives.) To view digital video files from the hard drive, you must switch PowerDVD to File mode.

Follow these steps to switch to File mode and to load files from your hard drive (or other source, such as a ZIP disk or a CD-ROM):

1. Click the Disc/File Mode button (Figure 8.2) or press O to toggle the button to File mode (the button graphic changes to a file folder).

2. Click the Menu/Playlist button to display the Edit Playlist window you see in Figure 8.10.

3. Use the tree display at the left side of the window to locate the digital video files you want to play—click on the plus sign next to a drive or folder to expand it. When you've located a digital video file that the program can play, it will appear in the Path list at the top right of the window.

4. Click the desired video file in the Path list and click Add to add it to the Playlist at the bottom right of the window. To add more video files to the playlist, repeat steps 3 and 4—PowerDVD will play them in the order they appear in the playlist.

5. Once you're done with the playlist, click OK, and click the Play button on the control panel to start the show.

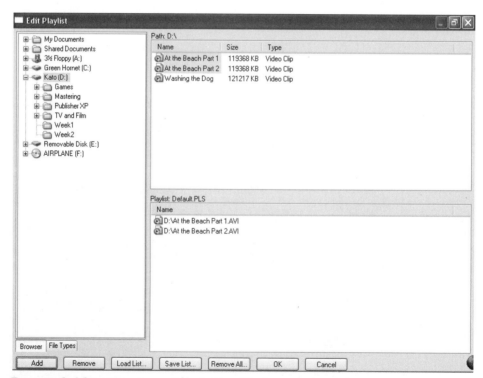

FIGURE 8.10 Choosing a digital video file from my hard drive.

Using the Menu Pad

As I've mentioned earlier, most DVD movies display an interactive menu that you can use to select different features—you can simply click your mouse pointer on a menu option to select it, but PowerDVD also provides a slick menu pad that simulates the keyboard pad on a standard DVD player remote control. Because some DVD movie menus may use "hotspot" buttons that aren't easily found on the screen, you can save yourself the trouble of hunting for those hotspots and use the menu pad to navigate the menu, instead.

To display or hide the menu pad, click the vertical button at the far right end of the control panel—you can also press the slash (/) key. Figure 8.11 illustrates the extended menu pad. Click on the desired directional button to navigate around the DVD movie's menu system (the button at the center of the pad is the Enter button, which usually activates the active menu command).

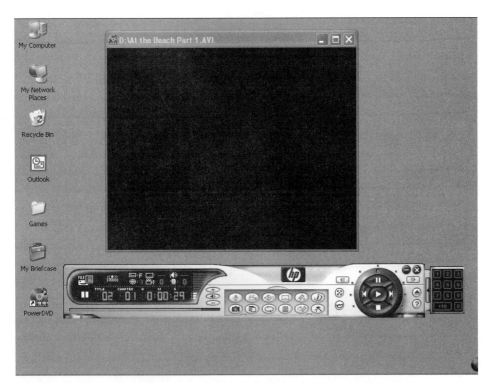

FIGURE 8.11 Using the menu pad to select a DVD movie menu option.

Project: **Capturing and E-mailing Images from Digital Video**

So you've used your new digital camcorder to film your daughter's first birthday, and some of the footage is priceless—I know the feeling! Let's assume that you want to e-mail selected images from the video to other family members, *without* sending them the entire 400-MB MPEG file. PowerDVD comes to the rescue, allowing you to capture images directly from your digital video and save those images as still photos—which you can then send through e-mail as attachments.

Because you may not be familiar with e-mail attachments, this project will take you all the way from capturing the image to sending the message; I'll be using Outlook Express to send the images.

Requirements

- A digital video file on your hard drive (or a DVD movie disc)
- Outlook Express or other e-mail program

✔ **Follow these steps:**

1. Run PowerDVD—if the program is in Disc mode, click the Disc/File Mode button (Figure 8.2) or press O to switch to File mode.

2. Click the Menu/Playlist button to display the Edit Playlist window.

3. I'll navigate to the digital video file "Family Birthday"—because PowerDVD recognizes the format, it appears in the Path list. Click the entry for Family Birthday in the Path list and click Add to add it to the Playlist.

4. Click OK to return to the PowerDVD control panel, then click Play to begin watching the video.

5. When you reach the scene that you want to capture as a still image, click Pause, then right-click the Capture button on the control panel

(Figure 8.12). Choose Capture to File—by default, this saves the image in the My Pictures folder in your my Documents folder under Windows XP. You can take as many images as the free space on your hard drive allows; PowerDVD names the first image PDVD_000.BMP and increments the number as you save successive photos.

Capture   Configuration

FIGURE 8.12   The Capture and Configuration buttons on the PowerDVD control panel.

tip **To save the image to another directory, click the Configuration button on the control panel (Figure 8.12), and click the Capture tab to display the settings you see in Figure 8.13. Click Browse and navigate to the directory where you want your images saved, then click OK to save your changes, and click OK to exit the Configuration dialog.**

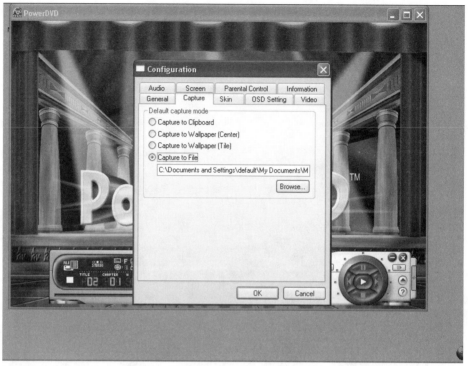

FIGURE 8.13   Changing the default Capture settings.

tip PowerDVD saves your still images in Windows bitmap format, which (depending on the size of the video frame) may be too large to send as an e-mail attachment—in general, I never send a message with more than a total of about 500 KB attached to it, because many Internet Service Providers will not allow large attachments. (You'll know that this is the case if you receive an error message from your e-mail server saying that the message couldn't be delivered or was too large.) However, there are two solutions to this problem of attachment "elbowroom": you can use a ZIP archiving utility, such as WinZIP (www.winzip.com), to reduce the file size, or you can load the image into an image editing program, such as Paint Shop Pro (www.jasc.com) or Photoshop, and convert it to a JPEG image.

6. Now that you have your images saved to your hard drive, it's time to send that e-mail. Run your e-mail program and compose a new

message; in Outlook Express, for example, you would click the Create Mail button on the toolbar.

7. Fill out the To and Subject fields as you normally would, and enter the text of your message. In Outlook Express, click the Attachment button on the New Message dialog toolbar to attach one of your images to the message.

8. Navigate to the location of the image files and click on one or more of the desired filenames, then click Attach. As you can see in Figure 8.14, the files appear in the Attach field. (Note that the file sizes are listed, too—remember, try to keep the total size of all attachments for a single message under 500 KB. If you need to, you can always send multiple messages with one image attached to each message. If the recipient is using a dial-up modem connection, it's a good idea to send the smallest attachments possible!)

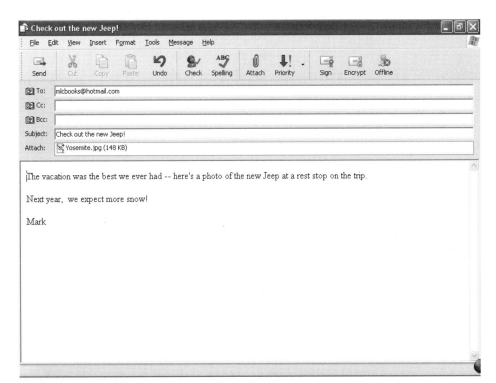

FIGURE 8.14 I've attached an image to this e-mail message.

9. Click the Send button in the toolbar to send the message.

10. If you've sent your images in Windows bitmap or JPEG formats and the recipients are running Windows ME or XP, they should be able to simply double-click on the icon for each attachment when they open the message. Figure 8.15 illustrates our attached image after I've opened it in Microsoft Outlook.

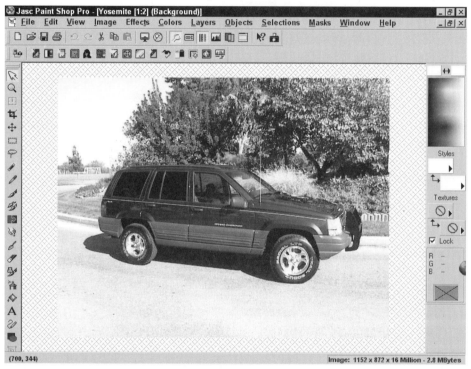

FIGURE 8.15 Voilà! The image from your digital video has crossed the Internet via e-mail.

Summary

In this chapter, we covered the features of PowerDVD, and you learned how to view DVD movies and digital video clips using this powerful application. I also demonstrated how to capture high-quality still images from DVD movies and digital video clips.

9

Printing Disc Labels and Inserts

In This Chapter

✔ Determining whether a disc needs a label or inserts

✔ Applying labels the right way

✔ Designing inserts

✔ Project: Creating a CD Label

✔ Project: Creating a CD Case Cover and Insert Set

Nothing adds a professional appearance to a CD you've recorded better than a custom label—except, perhaps, a custom label and case inserts! As long as you label your recorded discs in some fashion and protect them somehow, you've taken care of the basics; however, a felt-tipped permanent marker and an empty compact disc "jewel box" case won't make a great impression on anyone.

In this chapter, I'll demonstrate how you can use Roxio Easy CD Creator and your computer's printer to turn out CD labels and case inserts that are both attractive and informative—in addition, I'll show you how to apply a custom CD label properly.

Do I Need Labels and Inserts?

If you're recording only DVD+RW discs, you should think twice before applying a paper label. For example, Hewlett-Packard includes this warning with each HP DVD+RW disc:

caution   **Never use a ball point pen or adhesive labels on DVD+RW media—there's too high a risk of damaging the disc! Any change you make to the balance of the disc (no matter how slight) will likely render a disc unreadable. Remember, DVD technology packs more data on the same disc because the track containing the pits and lands is far smaller, so there's less room for error.**

Therefore, this chapter applies only to those folks who are using their DVD recorder to burn data and audio CDs using CD-R and CD-RW discs, as well as those who want to create custom case inserts for their DVD cases.

With that said, when should you take the time to create custom labels and inserts? I recommend them when:

- You're recording discs for distribution—for example, if you're a shareware author or a musician recording demo discs. A label is also a good idea if you're sending beta copies of software to testers or distributing a disc among your customers.
- You're presenting a disc as a gift.
- You'd like to document the contents of a data or audio CD fully. Because you can add a surprising amount of text to both the label and case inserts, you can even include luxuries such as track times and liner notes.

tip Labels are a great idea when creating hard drive backup sets using CD-RW discs—with a label, you'll have more space to add dates and hard drive information.

- You simply want the best appearance for your collection of recorded discs.

Naturally, I don't create disc labels for every project—but when you take the time to create labels, they can really make an impression! (And add convenience, as well; for example, all of my home movie DVD discs have custom case inserts with thumbnail images from the video clips. I can find the right disc in seconds.)

Applying Labels 101

Two different types of CD labels are available: those that are designed to be applied by hand and those that are designed for application using a device such as the two-piece NEATO CD labeling machine in Figure 9.1 (www.neato.com).

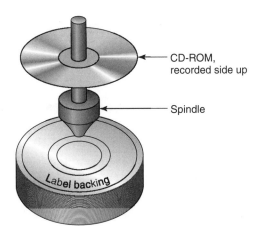

CD-ROM, recorded side up

Spindle

Label backing

FIGURE 9.1 Applying a CD label using the NEATO CD machine.

If you choose labels that are applied by hand, leave the disc inside its jewel case; this helps steady the disc and allows better alignment. In fact, most of these labels include a border that acts as a visual guide—you match it to the edge of the case. With hand-applied labels, it's important to smooth the label slowly across the face of the disc, eliminating bubbles as you go.

caution   **Never attempt to remove a paper label from a disc unless the manufacturer specifically calls the labels "removable"! Most CD labels use permanent adhesive, and you're likely to damage the disc beyond repair; in addition, a badly-aligned label can throw your disc out of balance and ruin it. (Yet another reason to be careful when aligning and smoothing a disc.)**

The NEATO device (which dates back to the mid-1990s) is a little more hi-tech, and it guarantees correct alignment with very few bubbles: The disc is placed upside-down on the spindle with the reflective side facing up, and the printed label is placed on the base with the adhesive side facing up. The label is applied by pushing the spindle into the base and pressing down.

Which type of labels is best for you? Personally, I think the labeling device is worth the extra investment if you need to label your discs quickly or if you produce a large number of discs at one time. A label applicator is a must for the best quality of work.

tip   **When peeling the label off the backing sheet, here's a trick that will prevent the label from "rolling up" into a tube: Turn the sheet upside down and lift up a small area of the edge of the label, then hold that down against a table. Pull the sheet up and off the label, and you'll be left with the label on the table sticky side up and flat as a pancake!**

Project: **Creating a Label**

Enough talk: Let's print a custom label for an audio CD! In this project, you've created a "Greatest Hits" birthday collection from a friend's favorite CDs, and you want to present it as a gift so it can be played during the party. I'll be using CD Label Creator, which is included in the popular Easy CD Creator recording software from Roxio (www.roxio.com).

Requirements

- Computer with a laser or inkjet printer
- Installed copy of Easy CD Creator 5
- Recorded audio CD

✔ **Follow these steps to design and print your label:**

1. Click Start | Programs | Easy CD Creator 5 | Applications | CD Label Creator to run CD Label Creator. Click the CD Label icon to select it at the left side of the screen, which displays the blank layout shown in Figure 9.2.

2. If you recorded your audio disc with CD-Text, you can import the title and artist name automatically—click the Contents toolbar button. (If you didn't use CD-Text or the program can't find the disc in the Internet CD database, you'll have to do things the old-fashioned way: Double-click in each of the fields and type the information manually. Press Enter to update the field once you've entered the text.)

3. Click the Theme toolbar button to select one of Roxio's preset themes. The Change Theme dialog appears, as illustrated in Figure 9.3.

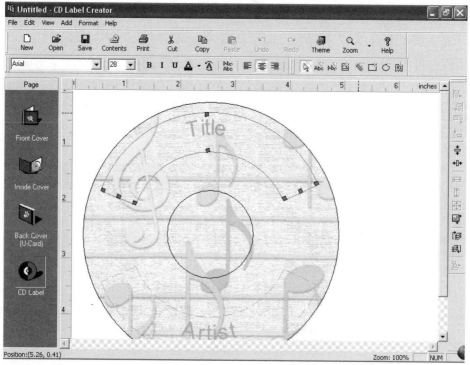

FIGURE 9.2   Roxio's CD Label Creator, ready for action.

4. Note that each of the themes listed in the Available Themes list is followed by either "Audio" or "Data"—this indicates which type of disc is preferred for use with that theme. (Any theme can be used for either type of disc, but these tags indicate which type of information will be automatically updated. If you use an audio CD theme with a data CD-ROM, you'll have to do more work and create all of the fields manually.) For our festive party, let's choose the Whirl theme—click Whirl in the list, which updates the Preview pane, and click OK. Our CD label layout now looks like the one you see in Figure 9.4.

tip **To view just the themes for an audio CD or a data CD-ROM, use the Show Audio Themes and Show Data Themes checkboxes at the top of the dialog box.**

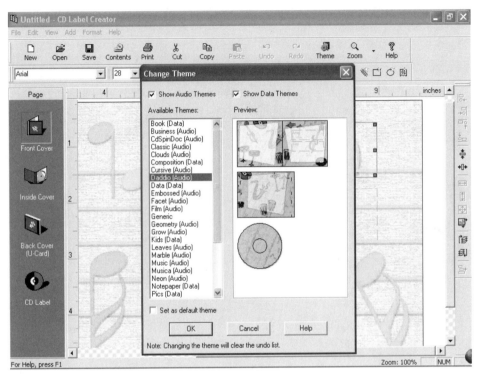

FIGURE 9.3   Selecting a new theme for our CD label.

5. CD Label Creator uses the rusty, tried-and-true default font of Times New Roman—boring! Because you can make changes to the text box that's currently selected, let's choose a new font and make things a bit more interesting visually. First, click on the Title box to select it, then click on the Font drop-down list box and click on a font name. To change the font size, click on the Points drop-down list box (the program automatically wraps words that are too big for the text box).

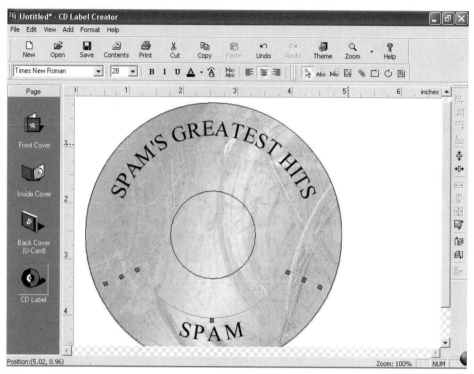

FIGURE 9.4   The Whirl theme has been applied to our CD label layout.

6.   Click on the Artist text box and choose the same font and point
     size. Figure 9.5 shows our disc sporting text using the Mickey
     font, which is much more fitting for a party.

7.   Our gift needs a little personalization. To add a text box, click Add
     and choose Text. Click and drag to move the text box below the
     spindle hole, as shown in Figure 9.6. You can also click and drag
     the *handles*—those small squares on the border of the text box—
     to resize the box.

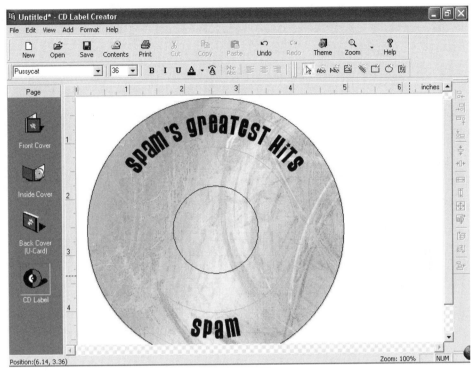

FIGURE 9.5   A change of font can do wonders for a disc label.

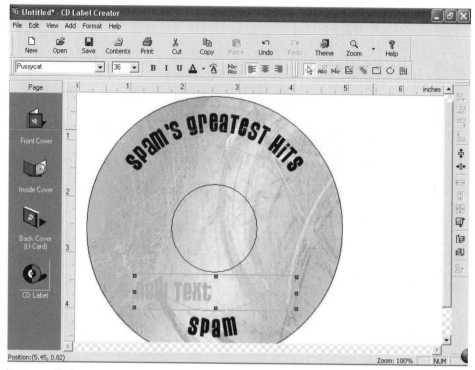

FIGURE 9.6 Adding a new text box to the layout.

8. Double-click in the box and type the text, then press Enter to save your changes. Figure 9.7 illustrates our completed label design, ready to print or save to your hard drive.

9. Load your CD label blanks into your printer as instructed by the manufacturer—make sure the labels are facing in the right direction, and the correct edge is facing away from the printer.

10. Click File and choose Print Preview (Figure 9.8) to check the appearance of the label. To make changes, click Close and return to the layout.

11. Click Print on the toolbar to print your labels.

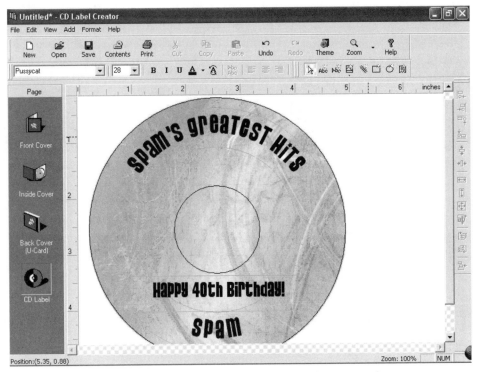

FIGURE 9.7   A completed label design can be saved or printed.

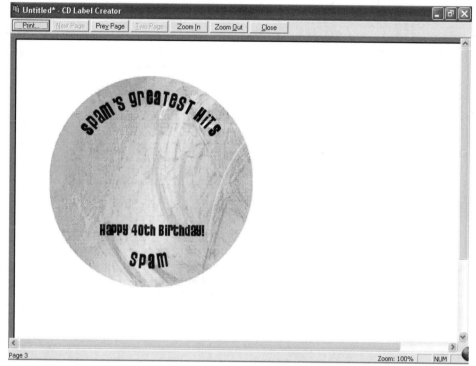

FIGURE 9.8   Why waste expensive labels? Use Print Preview to check your design.

Project: **Creating an Insert Set**

Now let's turn to another fun project that's easy to print with CD Label Creator: this time, a set of inserts for a DVD home video disc that I've just recorded. I want to use thumbnails from the video as artwork, and each clip will be noted on the back.

Requirements

- Computer with a laser or inkjet printer
- Installed copy of Easy CD Creator 5

✔ Follow these steps:

1. Click Start | Programs | Easy CD Creator 5 | Applications | CD Label Creator. Click the Front Cover icon to display the layout you see in Figure 9.9.

2. Let's select a new theme—the default, Music, doesn't work well for our DVD inserts. Click Format and choose the Change Theme menu item.

3. Click on the Pics entry to select the Pics theme, as shown in Figure 9.10—CD Label Creator automatically updates the Preview pane so you can see the theme—and click OK to accept it.

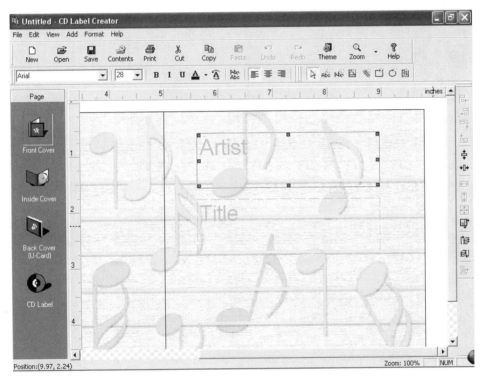

FIGURE 9.9   A new blank front cover layout.

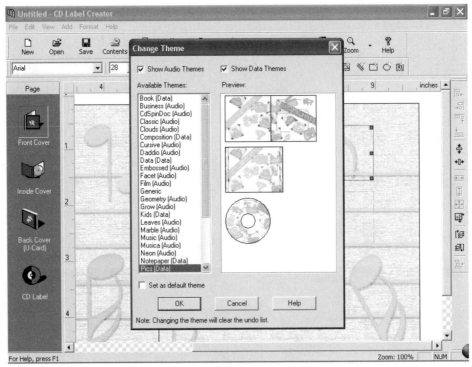

FIGURE 9.10   The Pics theme is perfect for a home video DVD.

4. Double-click directly on the Title field. An edit box appears, and you can type the title directly in the field. Press Enter to save your changes.

 tip **Made a mistake while editing your layout? Use the Undo button on the toolbar to backtrack from your last action.**

5. Double-click in the Contents field and enter the name of each video clip on the disc—I usually add the date each clip was taken, as well as the subjects. Press Enter to save your changes. Your layout should now look something like mine in Figure 9.11.

6. Let's add a piece of clip art—in this case, a thumbnail image from the video. Click Add and choose the Picture menu item to display a standard File Open dialog. Navigate to the location of the image,

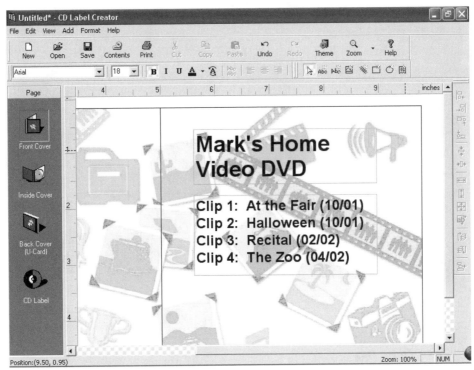

FIGURE 9.11 The completed text fields for my DVD home video disc.

click on the filename to highlight it, and click Open. CD Label Creator can use both Windows bitmap and JPEG image formats.

7. As you can see in Figure 9.12, the image is already selected. You can resize it and move it just like a text box. To resize the image, click on one of the square "handles" on the edge of the window and drag to resize that edge. To move the entire image to another spot on your layout, click in the middle of the selected image and drag it to the desired spot.

8. Don't forget the inside cover and back cover! Click on the desired icon in the Page view on the left side of the screen, and you'll find that you have quite a bit of additional space that you can fill with text and graphics. For this project, I'd like to add a quick description of each of my video clips on the back cover, so I'll click the Back Cover (U-Card) icon. CD Label Creator switches to the Back Cover

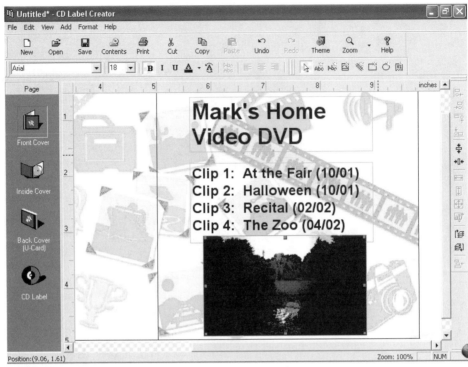

FIGURE 9.12   Adding a thumbnail image to my DVD insert layout.

layout you see in Figure 9.13—note that some of the information from the front cover has already been automatically imported.

9. To add a new block of text on the back cover, click Add and choose Text or Curved Text. A text block appears, and you can resize and relocate it as usual. To enter text, double-click in the desired text box and type; press Enter to save your changes. Figure 9.14 illustrates my completed back cover.

10. Ready to print? You can use either plain paper or precut insert paper—if you have insert paper, load it into the printer as instructed by the manufacturer.

11. To check your work before you print, click File and choose Print Preview. If everything looks good, click Print on the toolbar; to make changes or add more information to the inserts, click Close and return to the layout.

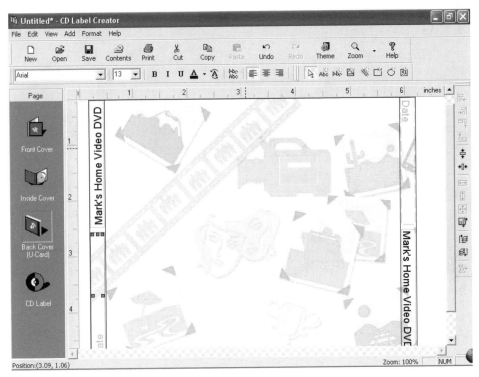

FIGURE 9.13 Adding a back cover to my insert set.

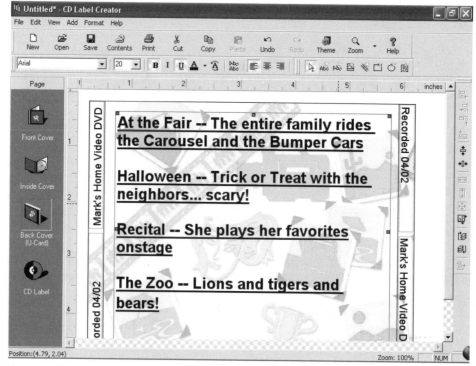

FIGURE 9.14 The back cover insert, ready to print.

Summary

We covered the world of CD labels and inserts in this chapter: when you should use them, how to apply them, and how to design and print them, using Roxio's CD Label Creator.

chapter

10

Recording Advanced Formats and Video CD Discs

In This Chapter

✓ Comparing Video CDs and DVD Video discs

✓ Understanding mixed-mode recording

✓ Understanding multisession recording

✓ Recording with a disc image

✓ Project: Creating a Video CD with Digital Video

✓ Project: Recording a Promotional CD-Extra Disc

✓ Project: Burning a Bootable CD

Throughout most of the earlier chapters, I've focused on recording standard DVD data and video formats, as well as basic data and audio CDs. However, there's an entire chapter's worth of advanced CD recording formats that have been popular for many years that I need to discuss— you might not need one of these specialized disc formats every day, but it's important to know how to create video CD, mixed-mode, bootable, and multisession discs when the situation calls for them.

y application of choice for this chapter is Roxio's Easy CD Creator Platinum, which ships with many CD and DVD recorders; you'll also learn how to create copies of a CD or DVD using a disc image. Our three projects will demonstrate how to record a standard Video CD disc, a CD-Extra disc that can be used with DVD players and audio CD players, and a bootable data CD.

Note that MyDVD version 3.5 will also create a proprietary type of Video CD disc called a cDVD disc; for more information on burning cDVDs with MyDVD, see the section titled "Recording the Disc" in Chapter 6.

Introducing Easy CD Creator 5

Figure 10.1 illustrates the main window from an old friend of mine: version 5 of Roxio's Easy CD Creator Platinum, which I've been using for about six years now. (Coincidentally, I also wrote my first book on the topic of CD recording in 1997. I'll bet you can guess which program I covered in that book.)

Easy CD Creator is a complete recording application—the Platinum version includes a number of standalone utilities that can produce:

- CD labels and jewel box inserts (as you saw in Chapter 9, "Printing Disc Labels and Inserts")
- UDF (packet-writing) discs using DirectCD, which operates very much like HP DLA
- Digital photograph albums and video postcards (more on this in Chapter 11, "Recording Digital Photographs on CD")
- Audio CDs from cassettes and albums, using your home stereo
- Hard drive backups using CD and DVD discs (such as HP Simple Backup)

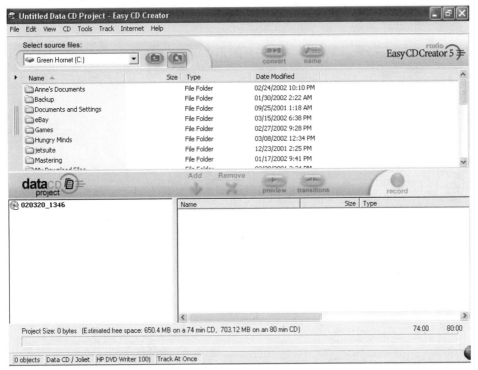

FIGURE 10.1 The Easy CD Creator main application window.

Although you can configure a disc format manually if you like, Easy CD Creator comes with a great "front-end" wizard program that can help automate the setup for many types of discs: We'll use the Roxio Project Selector (shown in Figure 10.2) often during this chapter. The Project Selector automatically starts Easy CD Creator with the proper settings for the standard formats you've learned about throughout earlier chapters.

If you've installed Easy CD Creator Platinum already, you can run the Project Selector in Windows XP by clicking Start | All Programs | Roxio Easy CD Creator 5 | Project Selector.

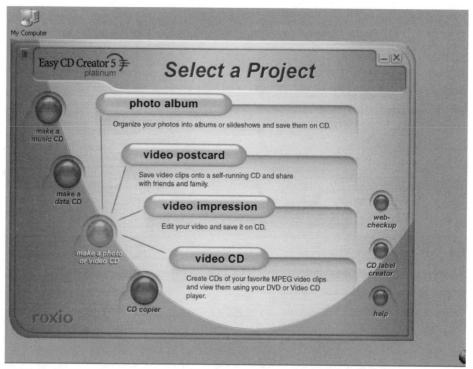

FIGURE 10.2   The Project Selector shows off the different video and photo discs it can create.

Video CDs vs. DVD Video Discs

"Mark, why do I need to record a video CD, anyway? Aren't DVD Video discs far superior to anything that's come before?" Good question, and the answer is a definite "yes"—DVD video discs are the best medium for carrying high-quality, high-resolution digital video.

However, not everyone has a DVD player or a DVD-ROM drive, and that's where the video CD still comes in handy! A video CD can't carry anywhere near the same amount of digital video, and the quality is significantly lower than a DVD Video disc—however, most computers with a standard CD-ROM drive can read a video CD, and many DVD players are also compatible with video CDs. Before the advent of DVD, many manufacturers turned out video CD players and CD-I players that will accept this format, as well. As you learned in the section titled "Loading and Playing a DVD Movie" in Chapter 8, PowerDVD can display video CDs on your computer.

Therefore, here's one of Mark's Patented Recording Rules: *Burn DVD Video discs when you're sure a DVD player or DVD-ROM drive (that you know is compatible) will be handy, and burn video CDs for friends and family who aren't prepared for DVD yet!*

Depending on the compression scheme you're using, you can usually fit anywhere from 45 to 75 minutes of MPEG video on the standard video CDs produced by Easy CD Creator. Note that I said MPEG video—in order to record video clips in Windows AVI or QuickTime MOV formats, you'll have to convert them to MPEG format first. (In some cases, Easy CD Creator can do this for you automatically.) A video CD can even include a rudimentary menu system that will allow you to select the clip you want to watch (when used on compatible DVD, video CD, and CD-I players).

Introducing Mixed-Mode Recording

In earlier chapters, you learned the differences between recording audio and data CDs—oil and vinegar, right? Never will the two mix. (I love this part.) Unless, of course, you move to *mixed-mode* recording! A mixed-mode CD-ROM actually has both digital audio tracks and a data track on the same disc; a good example is a multimedia CD-ROM game title that includes digital video, a computer application, and several tracks of digital audio. Just like English in elementary school: There's an exception to every rule!

To illustrate how this magic is performed, check out Figure 10.3: You'll note that the first track on the disc is computer data, and the following tracks are recorded separately as digital audio.

Naturally, an idea this good doesn't come without caveats. Before we get too excited about mixed-mode recording, here's the downside:

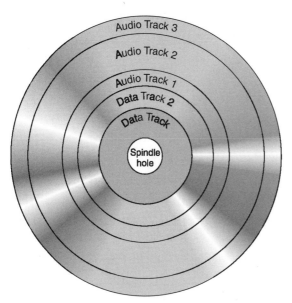

FIGURE 10.3 Merging computer data and digital audio on the same mixed-mode disc.

- **Older hardware may not recognize a mixed-mode disc.** If an older PC is using a CD-ROM drive that's 4X or slower (or if you scavenged such a drive from a garage sale or an eBay auction), it probably won't be able to read either the data or the digital audio from a mixed-mode disc.

- **Older operating systems may not recognize mixed-mode discs.** Older versions of UNIX, Mac OS, and DOS won't be able to read a mixed-mode disc—luckily, these relics don't really have the horsepower to play digital video or high-quality audio anyway, so good riddance!

- **You can't play a standard mixed-mode disc in your audio CD player.** Remember that the first track of a standard mixed-mode disc is still computer data, so your audio CD player will spit it out (or play some incredibly bad audio). You can get around this problem by recording a very useful variant of the mixed-mode standard disc: The *CD-Extra* (also called *Plus* or *Enhanced*) disc shown in Figure 10.4 is recorded with the audio

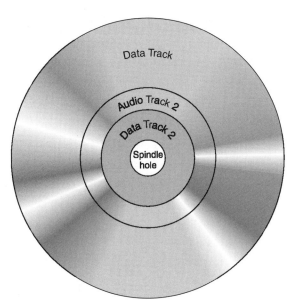

FIGURE 10.4 Your audio CD player accepts a CD-Extra mixed-mode CD-ROM without any trouble.

tracks appearing first on the disc, so it can played on a standard audio CD player. (Your audio CD player simply ignores the data track that appears at the end of the disc.) Bands such as the Rolling Stones and Smash Mouth have released CD-Extra discs that include the music, a music video, and all of the song lyrics on the same disc! We'll record a CD-Extra disc in one of the projects at the end of this chapter.

Introducing Multisession Recording

A traditional single-session data CD-ROM has only a single recording session—the disc is then fixed and closed so that it can't be recorded again. With a multisession disc, however, you can record it once, use it, then record on it again; for instance, you could record your tax returns on the same disc, "appending" them each year until you fill the entire disc.

In multisession recording, you can record separate sessions (think of them as separate volumes on the same disc), or you can "update" the same session with new information. Each session you record is separated from the previous section by an area of empty space (which can range anywhere from 14 to 30 MB, depending on its position on the disc surface). Because of this lost space, you never get the storage capacity from a multisession disc that you get from a single-session disc. It's a good idea to avoid recording a session that contains only a small amount of data, because you're guaranteed to lose at least 14 MB when you add another session!

You can record two different types of multisession CD-ROMs, using Easy CD Creator:

1. **Incremental.** An incremental multisession disc makes it easy to "update" previous data that you've recorded, because all the information that you've recorded within the previous session is available; as the name suggests, you can add files to that existing information. Easy CD Creator includes an Import Session function that enables you to read the contents of the previous session into your data CD layout. Incremental discs are fast to create, too, because the information in the previous session is not actually rerecorded. Instead, the session's directory information is retained and updated with the new data; however, your drive can read only the latest session you've recorded.

tip **You can even "delete" files from an incremental multisession disc! (Actually, files deleted from this type of disc are not erased and you don't regain any of the space they used; instead, the location of the file isn't updated in the disc's Table of Contents, so it effectively disappears and can't be recovered.) This trick was used often in the days before UDF (or packet-writing) became common.**

2. **Multivolume.** Data on a multivolume multisession disc is arranged in separate sessions, and you can switch between volumes whenever you like—however, you can read data from only one volume at a time, and you can't modify the data stored on a

session. Note that older, first-generation CD-ROM drives may not be able to read all of the sessions on a multivolume disc—instead, they may be able to access only the first or the last session on the disc. (Consider an incremental multisession disc if your CD-ROM will be read on older computers.) Figure 10.5 illustrates multivolume recording at work.

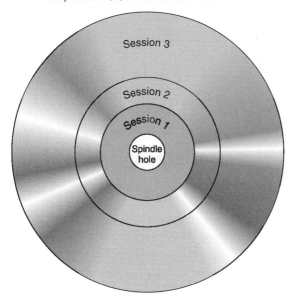

FIGURE 10.5 Separate sessions store discrete data in a multivolume multisession recording.

Recording an Incremental Multisession Disc

To burn an incremental multisession disc with Easy CD Creator, you need a disc with an existing session that you've recorded earlier. This disc must have been recorded with the Record Method set to *Track-at-Once* and *Finalize Session, Don't Finalize CD*. (These settings appear when you click the Options button on the Record CD Setup screen, as shown in Figure 10.6.) If you've created the first session on the disc with another recording program, make sure you don't *finalize* (or close) the disc, or you won't be able to write to it again!

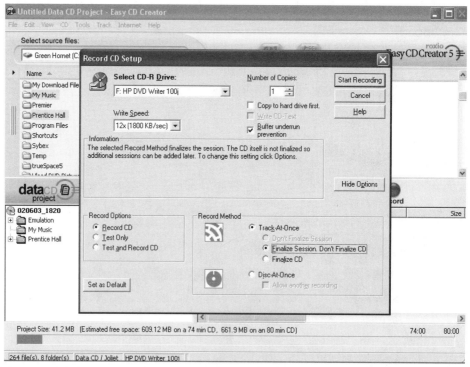

FIGURE 10.6 The proper recording options to use for incremental multisession recording.

✔ To record an incremental multisession disc:

1. Load the disc with the existing session into your drive. Easy CD Creator automatically displays the Roxio Project Selector.

2. Place your mouse pointer over Make a Data CD to display the different types of data disc you can record.

3. Click Data CD Project to load the Data CD Layout screen.

4. Before you add any files or folders to the empty layout, click the File menu and select CD Project Properties to display the dialog you see in Figure 10.7. Verify that the *Automatically import previous session check box is enabled*—if not, click the check box to turn the feature on—and click OK.

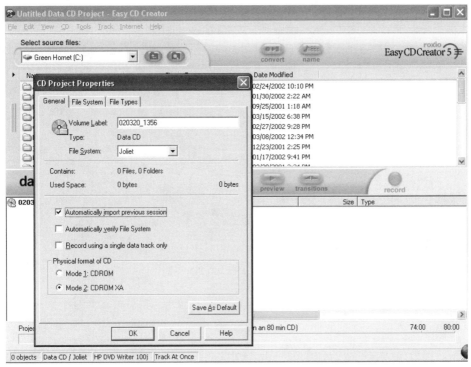

FIGURE 10.7 Setting Easy CD Creator to automatically import our existing session.

5. Build the layout for the new session: Navigate to the files and folders you want to add in the top portion of the window, highlight them, and click the Add button in the middle of the window to move them to the layout below. When you add the first new file or folder, Easy CD Creator automatically imports the files and folders recorded during the previous session.

tip **Things are very similar to Windows Explorer in the world of Easy CD Creator: For example, you can double-click on a file or folder name in the layout portion of the window to rename it, and you can drag and drop files in the layout to rearrange them as necessary.**

6. Click the big red Record button at the center of the screen to display the Record CD Setup dialog. If you'll be adding more data to

the disc later—in other words, if this isn't the last incremental session you'll add to the disc—remember to once again set the Record Method to *Track-at-Once* and *Finalize Session, Don't Finalize CD.* (If this is the last session you'll record on this disc, you can set the options to *Track-at-Once* and *Finalize CD.*) Click Start Recording to burn, and the progress dialog appears (Figure 10.8)!

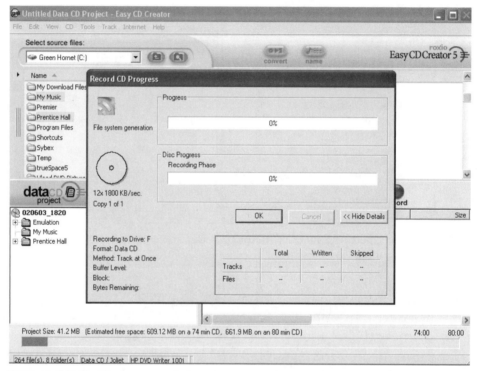

FIGURE 10.8 The progress dialog appears while recording an incremental multisession disc.

Recording a Multivolume Multisession Disc

To record a multivolume multisession disc, you'll once again need a disc with an existing session recorded with the Record Method set to *Track-at-Once* and *Finalize Session, Don't Finalize CD*.

✔ **Follow these steps:**

1. Load the disc with the existing session into your drive. Easy CD Creator automatically displays the Roxio Project Selector.

2. Place your mouse pointer over Make a Data CD to display the different types of data disc you can record.

3. Click Data CD Project to load the Data CD Layout screen.

4. Before you add any files, click the File menu and select CD Project Properties—verify that the *Automatically import previous session* check box is enabled. (If not, click the check box to turn the feature on.) Click OK.

5. Build the layout for the new session as I described in the previous section.

6. Click the big red Record button at the center of the screen to display the Record CD Setup dialog. (If you'll be adding more data later, remember to once again set the Record Method to *Track-at-Once* and *Finalize Session, Don't Finalize CD*.) Click Start Recording.

7. Easy CD Creator prompts you to confirm that you want to add a new volume to the disc. Click Yes to begin the recording.

Selecting a Session from a Multivolume Multisession Disc

The Platinum version of Easy CD Creator also includes Session Selector, a useful little utility application for selecting and reading a different volume.

✔ **To choose a new session from a multivolume disc, follow these steps:**

1. Run Session Selector from the Start menu by clicking Start | All Programs | Roxio Easy CD Creator 5 | Applications | Session Selector. The program displays the main window shown in Figure 10.9.

2. Click the correct CD-ROM drive in the left pane of the window. Session Selector displays the sessions on the selected disc in the right portion of the window.

3. The current session being read is reported with a drive letter. To select another session, click the desired session in the list to highlight it, click Tools, and choose Activate Session.

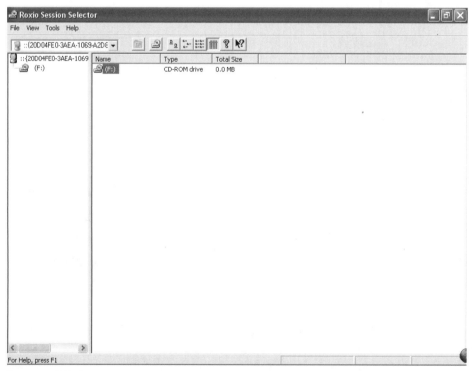

FIGURE 10.9   Using Session Selector to choose a session from a multivolume disc.

4. Click File and select the Close menu item to return to Windows.

tip **You can also select another session from the disc and automatically open it within Windows XP using Windows Explorer. To do this, click the session to highlight it, click the Tools menu, and choose Explore.**

Recording with a Disc Image

Consider a *disc image* as a single file on your hard drive that contains all of the information necessary to create a disc. There are several reasons why you may decide to create a disc image instead of actually recording a disc itself:

- **Security.** If you record a CD or DVD disc from a disc image, you can be sure of the stability of the disc (none of the files were altered or updated by mistake in the interim)—an important point for software developers, for example.
- **Speed.** If you have an older computer with a faster recorder that doesn't have burn-proof protection, you may experience recording errors (especially if you're recording a large number of small files) because of the slower data transfer rate. Recording your data as an image first will significantly reduce the chance of recording errors, because the image is a single continuous file; your hard drive doesn't have to read all those smaller files and send them to your recorder.
- **Convenience.** If you must record the same disc from time to time but don't want to create a batch of several copies at once, use a disc image; if the disc changes, you won't waste additional discs you recorded before the change.
- **Mobility.** Let me guess: no recorder on your laptop? (Or perhaps you're sharing an external CD recorder with others?) Record a disc image to your hard drive, and when the external drive is available, just record the disc from the image.

tip Make sure that you have at least 800 MB of free space on your hard drive for a full CD-R or CD-RW disc image—that's about 700 MB for the image itself, along with a little left over for Windows and Easy CD Creator to use.

As you can see, a disc image is useful in all sorts of situations!

Saving a Disc Image

✔ **To save a disc image to your hard drive, follow these steps:**

1. Run Easy CD Creator by clicking Start | All Programs | Roxio Easy CD Creator 5 | Applications | Easy CD Creator.

2. Click File and select New CD Project, then select Data CD from the pop-up menu. (This is another method of starting a new project.)

3. Build your CD layout.

4. From the File menu, choose the Create CD Hard Disk Image menu item. Easy CD Creator displays the familiar Image File dialog box shown in Figure 10.10.

5. Navigate to the location on your drive where you want to store the disc image, and type a filename. (The CIF extension is automatically added by Easy CD Creator.)

6. Click Save.

7. Take a soda or coffee break while Easy CD Creator "records" the image to the hard drive.

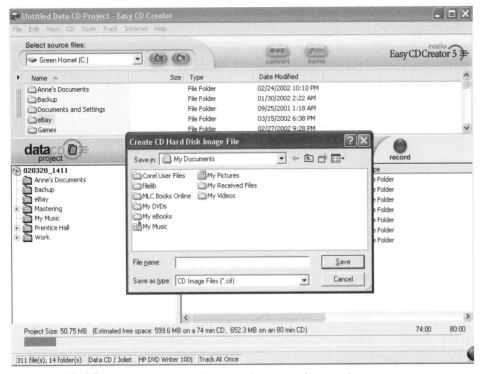

FIGURE 10.10 Selecting a location and filename for our disc image.

Recording a CD from a Disc Image

✔ **To record a disc from an existing disc image on your hard drive:**

1. Run Easy CD Creator.

2. Click File and select the Record CD from CD Image menu item.

3. Easy CD Creator displays the Image File dialog box.

4. Locate the disc image you want to record on your hard drive. Highlight it and click Open.

5. Easy CD Creator displays the Record CD Setup dialog you used in previous sections in this chapter. Click Start Recording.

6. Easy CD Creator prompts you to load a blank disc and records the disc.

Project: **Creating a Video CD**

In this project, we'll create a video CD using MPEG format digital video files you've recorded with your digital camcorder.

Requirements

- MPEG digital video clips
- Installed copy of Easy CD Creator 5
- A blank CD-R disc

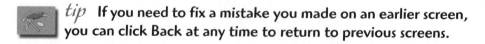

✔ **Follow these steps:**

1. Load a blank CD-R disc into your drive—avoid using CD-RW discs, because they're not compatible with most video CD hardware. Easy CD Creator will detect this and automatically display the Project Selector screen.

2. Move your mouse pointer over the Make a Photo or Video CD button, and click the Video CD button that appears.

   tip **If you need to fix a mistake you made on an earlier screen, you can click Back at any time to return to previous screens.**

3. Easy CD Creator loads the first wizard screen from the stand-alone program VCD Creator, as shown in Figure 10.11. Click Next to continue.

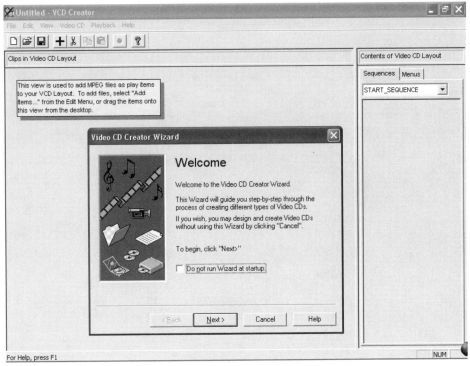

FIGURE 10.11 The Easy CD Creator VCD Creator wizard.

4. For this project, we'll use the Simple Video Sequence type—it provides the highest level of compatibility with DVD players and older CD-ROM drives. Choose Simple Video Sequence and click Next.

5. Click Add to select the MPEG video clips from your hard drive; VCD Creator displays the Add Play Items dialog shown in Figure 10.12. Navigate to the location of the MPEG files and double-click the first file to open it.

6. The clip is loaded and displayed in the Add New Play Item dialog, where you can watch the clip frame by frame, by dragging the slider bar under the preview window. Click OK to accept the clip. VCD Creator displays a thumbnail of the clip in the Video CD layout screen (Figure 10.13).

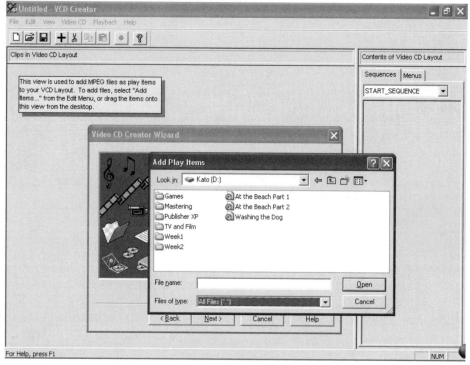

FIGURE 10.12 *Selecting an MPEG video clip.*

7. Repeat steps 5 and 6 as necessary to load each of the desired video clips into the VCD Creator layout screen.

8. Once you've added all of the clips, click Next to continue. Click Next again on the Creating Play Sequence screen.

9. It's time to set up the order (or play sequence) for the clips you've selected. Click the thumbnail image in the left column for the clip that should appear first and click Add (Figure 10.14). Continue this process until you've added all the clips in the order they should appear. If you need to remove a clip from the sequence you're building—or you want to remove the clip from the sequence and place it elsewhere—click the thumbnail in the START_SEQUENCE column to select it, and click Remove. Once you've added and arranged all of the clips in the order that

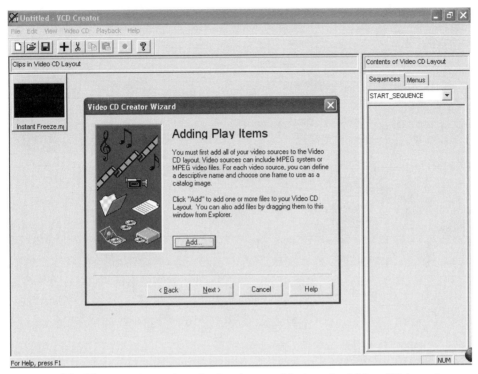

FIGURE 10.13 Our first video clip has been added to the Video CD layout.

you want them in the START_SEQUENCE column, click Next to continue.

10. Here's a nice feature: VCD Creator allows you to preview the video sequence you've created before you burn the disc! Click Playback, and VCD Creator displays the control panel shown in Figure 10.15—if you've used a VCR before, these controls should be quite familiar. You can use these to play, pause, or skip to the next or previous clip. Once you've finished the playback, click Close on the MPEG Playback panel, and click Next to continue.

11. Time to burn your disc—click Create the CD now on the final wizard screen (Figure 10.16), and click Finish to start the process.

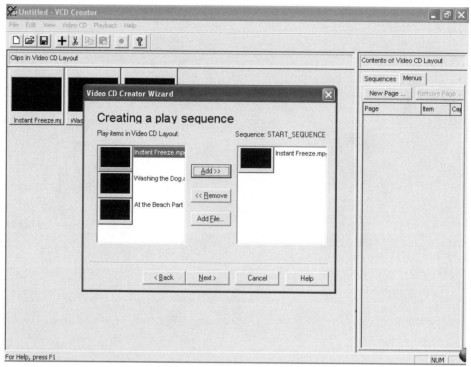

FIGURE 10.14 Specifying the sequence for my video CD clips.

Note that VCD Creator must create the video clip sequence file first, before the actual recording begins—depending on the speed of your PC, this can take a few minutes (especially if you've selected a large number of clips). When the video sequence is ready, you'll see the same familiar Record CD Setup dialog, and the rest of the process is the same as I've already described earlier in the chapter.

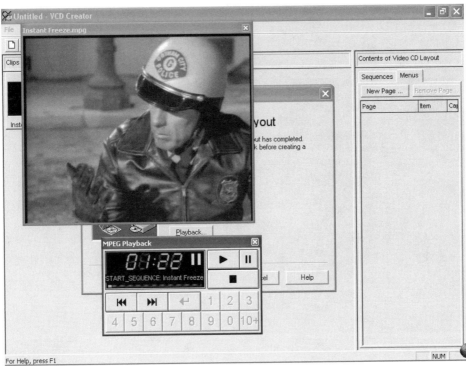

FIGURE 10.15   Previewing the video sequence, complete with VCR controls.

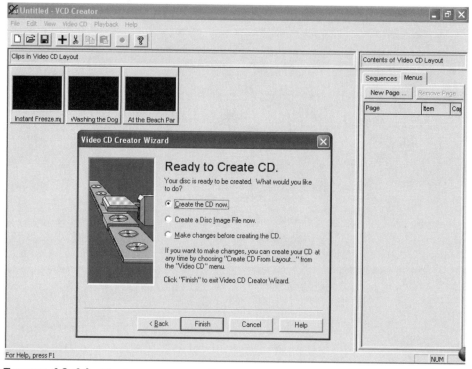

FIGURE 10.16   You're set to record!

Project: **Recording a Promotional CD-Extra Disc**

Here's a project for readers who've joined an up-and-coming garage band—I'll show you how to create a professional-quality CD-Extra demo disc for your up-and-coming rock band, with a number of audio tracks taken from MP3 files, followed by a data track containing the band's latest video and promotional information!

Requirements

- An MPEG, AVI, or MOV digital video clip
- A number of digital audio files
- Audio tracks taken from existing audio CDs

- Installed copy of Easy CD Creator 5
- A blank CD-R disc

✔ **To record music and video on a CD Extra disc, follow these steps:**

1. Run Easy CD Creator—from the Windows XP Start button, click Start | All Programs | Roxio Easy CD Creator 5 | Applications | Easy CD Creator.

2. Click the File menu, select New CD Project, and choose Enhanced CD (the other common name for a CD-Extra disc) to display the layout shown in Figure 10.17.

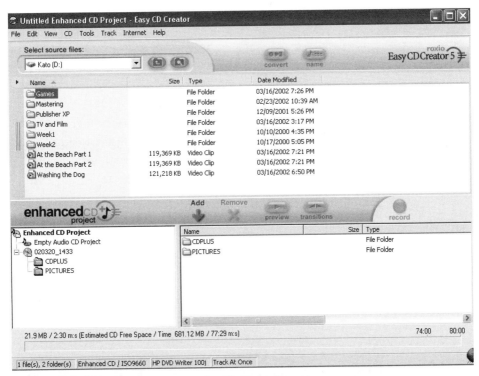

FIGURE 10.17 The beginning of a CD-Extra layout.

3. Click the Empty Audio CD Layout entry in the CD Layout window to add your audio tracks.

4. Navigate through your system and locate the audio files you want to record, using the Explorer display at the top of the window. You can also copy tracks from an existing audio CD by loading it and selecting that drive to display the tracks. We'll use five tracks in MP3 format, which Easy CD Creator will automatically convert during the recording process.

5. Click the audio files to highlight them. To select multiple files, hold down the Ctrl key while you click.

6. Click the Add button in the center of the window to copy the files.

7. Repeat steps 4, 5, and 6 until you've added all the tracks you want to the audio layout.

8. Next, it's time to build the data portion of the layout. Click the disc icon in the CD Layout window to add files and folders to the disc.

caution   **Don't delete the CDPLUS directory or any of the files it contains, and don't place any files or folders inside it! This folder is automatically added by Easy CD Creator specifically to hold any CD-Text information you enter for the audio tracks (as I will demonstrate in step 12).**

9. Navigate through your hard drive, using the Explorer display at the top of the Layout window, and locate the files and folders to record. Click the icon for the file or folder you want to add to your data CD to highlight it; hold down the Ctrl key while you click each icon to highlight multiple files and folders.

tip   **Although Easy CD Creator automatically assigns a default volume label for the data session, you can change it if you like by clicking the label and typing a new one.**

10. Click the Add button to add the files and folders to the data portion of the layout. In this case, I'm adding both a digital video clip and a promotional brochure about the band in Adobe PDF format.

11. Repeat steps 9 and 10 until you've added all the files you want to the layout—Figure 10.18 illustrates a completed CD-Extra layout.

12. If you like, you can enter additional information that can be displayed by audio CD players that support the full CD-Extra format: Click File, choose CD Layout Properties, and click the CD-Extra tab to display the dialog box shown in Figure 10.19. All fields in this dialog box are optional. Click the Created: and Published: drop-down list boxes to choose the desired date from a calendar pop-up display, and type text directly into the other fields. Once you're done, click OK to save the information to your layout.

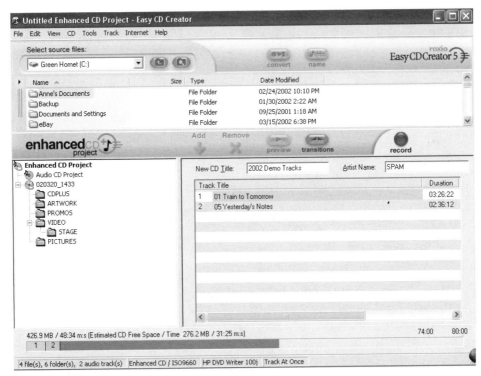

FIGURE 10.18 Our CD-Extra layout is complete.

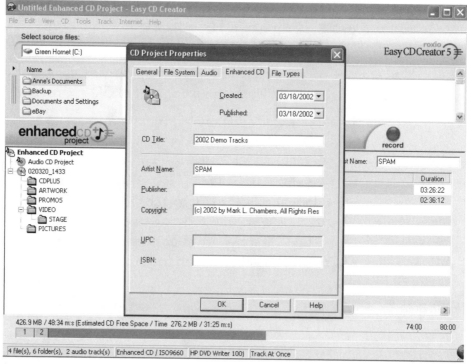

FIGURE 10.19 Adding audio CD text to our layout.

13. Click the big red Record button and complete the recording process, as demonstrated earlier in this chapter.

Project: **Recording a Bootable CD-ROM**

Most normal human beings will likely never need to actually record a *bootable* CD-ROM—a disc that contains the operating system files required to boot your computer—but if your hard drive fails or you're a computer technician, you'll find this specialized format invaluable! (Here's a piece of absolutely meaningless-but-fun trivia: Bootable CDs use the *El Torito* format standard. I don't know where that name came from, but it really adds sparkle to this discussion, don't you think?)

Anyway, most operating systems are shipped on bootable discs—for example, the Windows XP install disc is a bootable CD-ROM—and many Windows utilities are also designed to work from a bootable CD (such as Norton Utilities and Norton AntiVirus). If you bought your PC within the last three years or so, it probably supports a bootable disc—check your computer's manual or display the BIOS options the next time you turn on your PC, and look for "CD-ROM" as a startup boot drive option.

Because bootable CD-ROMs can hold over 600 MB, you can usually store all of the operating system files you need on one disc, as well as any hardware drivers you'll use, utility software, and troubleshooting programs.

caution   **As you might expect, it requires more expert technical knowledge to create an El Torito bootable CD than any other type of CD format I discuss; therefore, I recommend that you try this project only if you're an experienced PC technician or power user.**

To record a bootable CD, you'll need a bootable floppy "master" disk ready that contains all the files necessary during the boot process; check your operating system help files to determine how to format a bootable system disk. (You can also use an Easy CD Creator disc image that you built earlier.)

tip **If you use a floppy to create a bootable CD and you're running Windows 98/Me, XP, NT, or Windows 2000, you'll be limited to DOS as the operating system for a bootable CD-ROM (because the complete operating system won't fit on a single floppy). Therefore, you should create and include AUTOEXEC.BAT and CONFIG.SYS files on your floppy disk if you want to configure your system automatically during the boot process.**

Requirements

- A bootable floppy
- Additional programs and data files

- Installed copy of Easy CD Creator 5
- A blank CD-R disc

✔ **Follow these steps to record a bootable CD-ROM:**

1. Run Easy CD Creator—from the Windows XP Start button, click Start | All Programs | Roxio Easy CD Creator 5 | Applications | Easy CD Creator.

2. Click the File menu, select New CD Project, and choose Bootable CD to display the dialog shown in Figure 10.20.

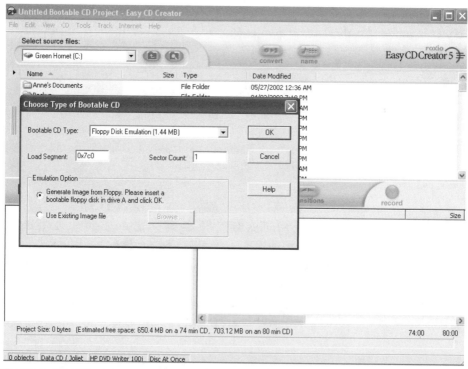

FIGURE 10.20 Selecting the options for a bootable CD.

3. Because we're creating a bootable CD from a DOS system disk, leave all of the settings as they are and load the bootable floppy disk into your floppy drive. Click OK.

4. The files are read from the floppy disk and stored in two special files within your CD layout: BOOTCAT.BIN and BOOTIMG.BIN.

5. Load a blank CD into your recorder.

6. From this point on, you can add any remaining files to your layout normally, just as you would for a standard data CD-ROM. You can create folders if necessary.

7. Once the layout is complete, click Create CD to record the disc.

Summary

You learned all about a number of advanced recording tricks and specialized CD formats in this chapter, including video CDs, disc images, bootable CD-ROMs, CD-Extra discs, and multisession recordings.

11

Recording Digital Photographs on CD

In This Chapter

✔ Creating a catalog of photographs using PhotoRelay

✔ Printing photographs

✔ Project: Creating a Family Photo Slideshow CD

✔ Project: Creating a Video Postcard CD

Although a recorded CD doesn't have anywhere near the capacity of a recorded DVD, there are still some applications that are better suited for CD-R or CD-RW—for example, most audio is still recorded on CD-R. CD-RW discs are also great for recording digital photographs, because today's typical digital camera produces a JPEG image of under 200 KB. With the file sizes for photographs so small, you can fit thousands of images on a standard CD-ROM—and you can be assured that any computer with a CD-ROM drive can display your photographs, too.

221

This chapter includes two step-by-step projects for storing both digital photographs and digital video on CD—the former as a slide show, and the latter as a neat "video postcard" that you can view on your computer. I'll be using a great program called PhotoRelay, from ArcSoft (www.arcsoft.com)—it's included as part of Easy CD Creator Platinum.

Introducing PhotoRelay

To run PhotoRelay, click Start | All Programs | Roxio Easy CD Creator 5 | Applications | PhotoRelay. The main program window appears, as shown in Figure 11.1. The PhotoRelay toolbar includes most of the common functions you're likely to need; to display the function of a toolbar icon, move your mouse pointer over the icon and check the status line at the bottom for its description.

PhotoRelay applies the familiar concept of a traditional paper photo album to your digital photographs: You create an *album* that contains pictures from your hard drive (either from a single folder or from different locations). When you display the contents of an album, each photograph is represented by a thumbnail image; this makes it easy to scan your pictures quickly for a particular shot. Once you've created the catalog, you can also:

- Attach audio clips to pictures in your album
- Print selected pictures or the entire album
- Sort the images in your album
- Build an automated slide show that can display your pictures
- Create an HTML photo catalog that you can add to your Web site
- Send a photo as an e-mail attachment

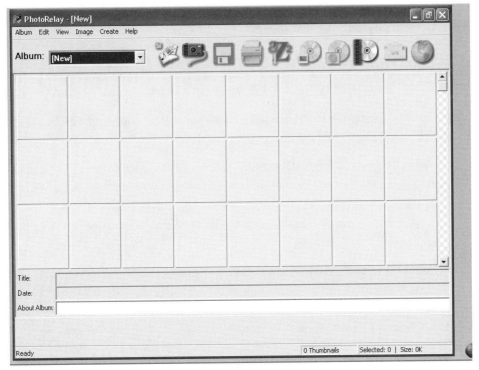

FIGURE 11.1 The main PhotoRelay window.

Creating a Thumbnail Album

Because the album is the starting point for everything in PhotoRelay, let's create a new album now. I'll add pictures from two sources: a folder of pictures that I've already downloaded from my digital camera and an image I'll acquire directly from my scanner.

> *tip* To download (or, in the technospeak of imaging gurus everywhere, to *acquire*) a picture from a scanner or digital camera directly into PhotoRelay, both your hardware and the software it uses must support the TWAIN standard. TWAIN is my absolutely favorite computer acronym: It stands for Technology Without an Interesting Name! (Never let it be said that computer techno-types don't have a sense of humor.) Anyway, virtually all scanners and digital cameras are TWAIN-compatible these days, but you must make sure

that your hardware's **TWAIN** drivers are correctly loaded before you can successfully acquire an image.

✔ Run PhotoRelay and follow these steps:

1. To add pictures from your hard drive, click the Add toolbar button—it's the first button on the toolbar.

2. PhotoRelay displays the Add to Album dialog shown in Figure 11.2; navigate to the folder that contains the first pictures you want to add. Click on the filename to select it (or, if you'd like to select more than one photograph, hold down the Ctrl key and click on each filename).

3. Click Open to add the pictures to the album file—Figure 11.3 illustrates a collection of 10 images I've added.

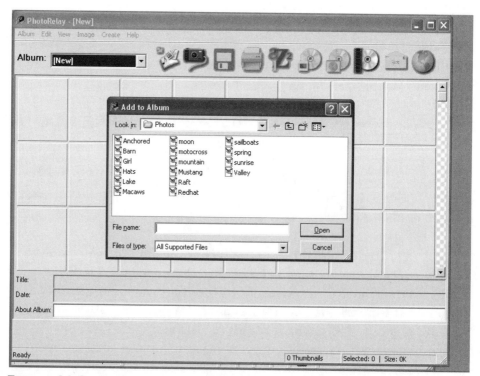

FIGURE 11.2 Adding images to an album from my hard drive.

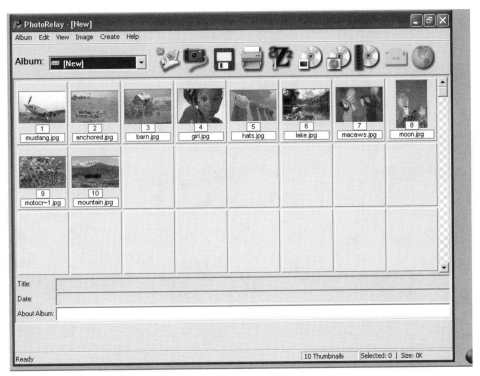

FIGURE 11.3   I've added a number of images from my hard drive to the new album.

4. Next, let's acquire an image from the scanner: Click the Acquire Images button on the toolbar (it's the second button, which looks like an animated camera). PhotoRelay displays the Select Source dialog you see in Figure 11.4, which lists all of the active TWAIN hardware on your system; in this case, I'll click my scanner. Click Select to continue.

5. At this point, your hardware will display its own acquisition dialog, so what you see won't look exactly like Figure 11.5—however, most scanners are likely to have most of the same controls. (If you're not familiar with your hardware's TWAIN acquisition settings, check the manual for more information.) Begin the scan—in my case, that means clicking Start Scan—and watch as PhotoRelay adds the image thumbnail to your album.

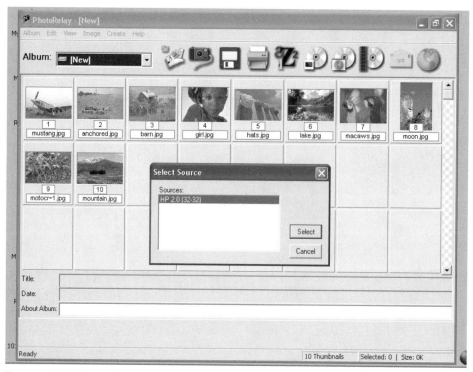

FIGURE 11.4 Specifying a TWAIN scanner as the source for new images.

6. Once you've added the photographs to your catalog, you can add a title, date, and description to any or all of the shots. Click on the desired thumbnail to highlight it, click in the desired field underneath the thumbnails, and type the corresponding text for the photo.

7. As a final step before saving the catalog, you can also sort the photographs in a number of different ways: Click the Album menu and choose the Sort menu item to display the Sort Album dialog you see in Figure 11.6. By default, PhotoRelay sorts the photographs in your catalog by their filenames; choose the desired sort order (note that you can enable the Reverse Order checkbox to switch between ascending and descending sort order), and click OK to rearrange your images.

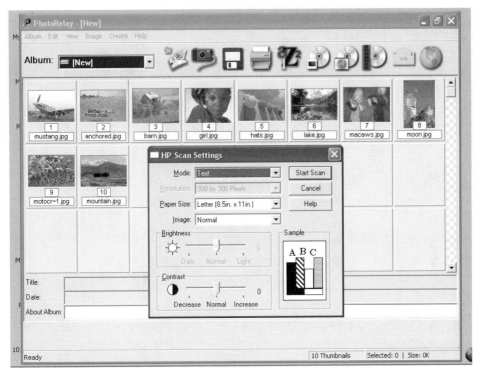

FIGURE 11.5 Configuring my scanner before acquiring an image.

8. Once you're satisfied with the contents, descriptions, and order of
   the photographs in your catalog, press Ctrl+S to save the album
   (or, if you're crazy about menus, click the Album menu and
   choose the Save Album item). PhotoRelay prompts you to name
   your new album, as illustrated in Figure 11.7 (note that the .PHB
   extension is automatically added); type the name, and click OK.

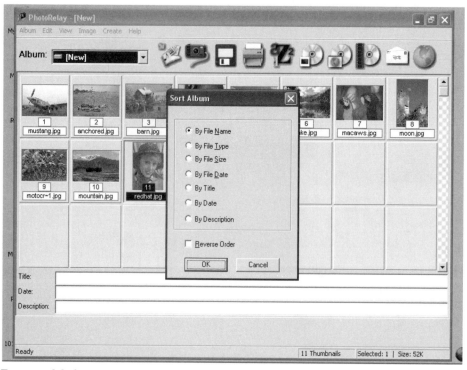

FIGURE 11.6 You've got a choice when sorting your pictures in PhotoRelay.

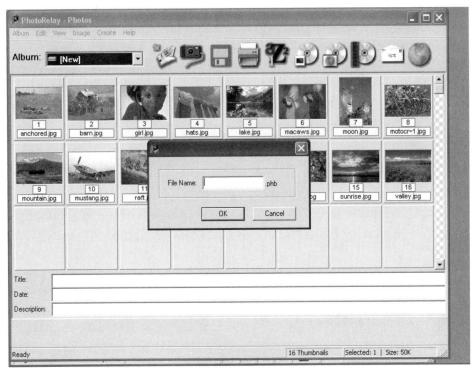

FIGURE 11.7 The final step: entering a filename and saving the album.

Printing Photographs

Although this chapter naturally focuses on recording your digital photographs, I'd like to take a moment to demonstrate how you can print a photograph from a PhotoRelay album. This program is wonderful for cataloging your images, and it's easy to print a photo directly from within an album—if you own a Hewlett-Packard photo-quality inkjet printer, for example, you can produce an image on glossy paper that looks as good as a print from a film camera!

✔ **To print a single photograph from within PhotoRelay, follow these steps:**

1. Click on the Album drop-down list to display the names of the albums you've created, and click on the desired album.

2. Click on the thumbnail for the photograph you want to print to select it.

3. Click the Print icon on the toolbar (you guessed it, the icon that looks like a printer), and PhotoRelay displays the Print dialog you see in Figure 11.8. To print a single image that you've previously selected, click Print the current selected image.

FIGURE 11.8 Will that be one photograph or many?

4. The second Print dialog (shown in Figure 11.9) allows you to add a title or frame—you can optionally center your photo in the middle of the page or even create a 5x5 jewel case label for a CD case. (A nifty feature, indeed!) Click the desired checkboxes to enable or disable these features as necessary.

5. Next, drag the slider on the right side of the dialog to change the size of the photograph on the printed page; PhotoRelay displays the effects of the resizing in the preview window. You can also click and drag the image placeholder around on the page to move it wherever you like.

6. If you elected to add a title to the page, double-click in the title placeholder to type the text; PhotoRelay displays the Edit Text dialog you see in Figure 11.10. Note that you can also highlight the title text in the edit box and choose a new font for the title.

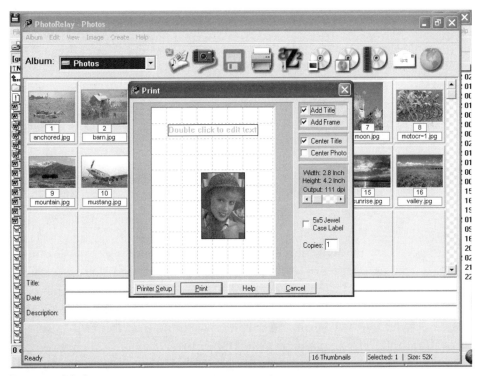

FIGURE 11.9 Setting options for our printed page.

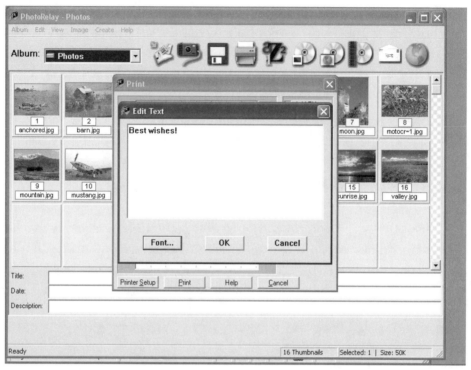

FIGURE 11.10   Changing the text for the page title.

7. By default, PhotoRelay prints a single copy of the photograph; to produce multiple copies of the picture, click in the Copies field and enter the desired number.

8. To print to your default Windows system printer, click Print—to choose another printer or to change printer options, such as DPI or image quality, click Printer Setup.

tip **PhotoRelay can also produce a very nice printed album of images, complete with a frame and background texture; you can also print text from your description fields, a title banner, and a custom footer line. If you want to print several pictures, hold down the Ctrl key while you click on the specific thumbnails. (No need to select any thumbnails if you want to print the entire album.) In step 3 of the printing process, click Print multiple photos or album—PhotoRelay will display a wizard that will lead you through the options. You can specify** all the pictures in the album, just those you've selected, or a numbered range of photos.

E-mailing Photographs

Besides printing images, PhotoRelay also works hand in hand with your e-mail software to send pictures from your albums across the Internet. By default, Windows XP uses Outlook Express for sending and receiving e-mail, so that's what I'll use in this section.

✔ **To send a single picture from within PhotoRelay using Outlook Express, follow these steps:**

1. Click on the Album drop-down list, and click on the desired album.

2. Click on the thumbnail for the photograph you want to send as an e-mail attachment to select it. (Note that you can select only one image when sending mail with PhotoRelay.)

tip When selecting a picture to send, don't forget that most Internet e-mail servers place a limit on the size of an e-mail attachment—whenever possible, it's a good rule to keep your attachments below 1 MB in size.

3. Click the SendMail icon on the toolbar—naturally, it looks like an envelope—and PhotoRelay automatically runs Outlook Express (or whatever e-mail program you've chosen in Windows) and prepares a new message, as shown in Figure 11.11. Note that the attachment has already been added; the size of the attached picture is also listed next to the filename, so you can keep an eye on the total file size for the message.

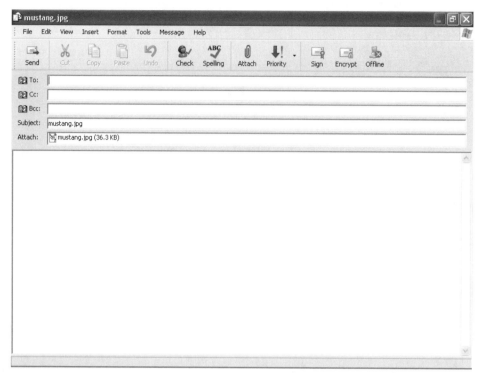

FIGURE 11.11 Sending an image in e-mail couldn't be much more convenient!

4. Type the recipient's e-mail address in the To: field (or click the To: button to add the recipients from your Address Book).

5. By default, PhotoRelay adds a subject using the name of the image, but you can click in the Subject: field and type a new subject if you like.

6. Click in the message edit box and type the text of your e-mail message.

7. When you're ready to "post" your e-mail message with the picture attached, click the Send button in the toolbar. Figure 11.12 illustrates a completed message, ready to send.

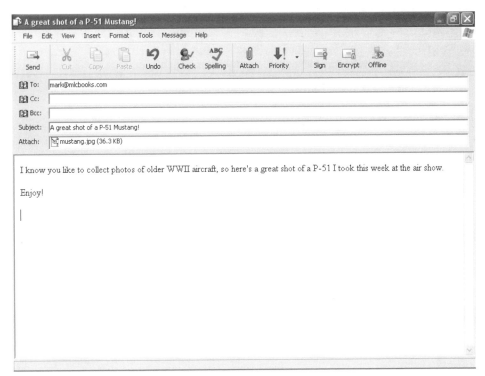

FIGURE 11.12 This message is ready to go.

Project: **Creating a Family Photo CD with PhotoRelay**

Ready to put your photographic memories in binary? Forget that old-fashioned slide projector…in this project, I'll show you how to create an impressive automated slide show using pictures from a photo album that you've already created. You'll be able to view the show on any PC running Windows 95 through Windows XP.

Requirements

- A PhotoRelay album of images
- A blank CD or CD-RW

✔ **Follow these steps to create your slide show disc:**

1. Run PhotoRelay.

2. Select the desired album from the Album drop-down list to load it.

3. Click the Create a Slide Show button on the toolbar (which looks like a CD-ROM, overlaid with a traditional film slide). PhotoRelay displays the Select File dialog box (Figure 11.13).

4. Hold down the Ctrl key and click each photo you want to add to the slide show in the left list box, then click Add. Note that, as you add pictures, PhotoRelay displays the amount of space on the CD required to create the disc underneath the right list box. Once you're done selecting the photographs for the slide show, click Next to continue.

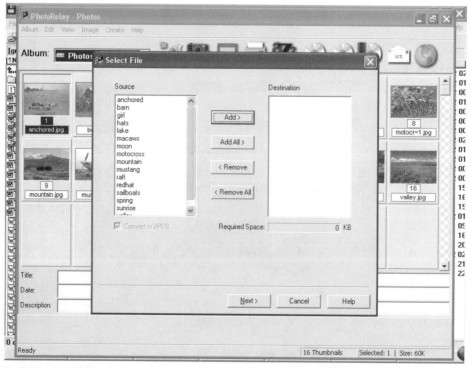

FIGURE 11.13 Selecting the pictures that will appear in our slide show.

tip **Do you want to add all of the photos from the entire album? Never mind clicking each photo name—to add all images in the album, just click Add All, then click Next to continue.**

5. From the Audio Options dialog (Figure 11.14), you can add a digital audio file in WAV or MP3 format to your show. The audio plays in the background while you're viewing your pictures. To use an audio file from your hard drive, click Play Single Audio File and click the Browse button to select the file, then click Open to load it. Click Next to continue to the next screen.

6. PhotoRelay displays the Select Destination dialog you see in Figure 11.15. Load a blank CD-R or CD-RW disc into your recorder. (Do not use a disc with an existing session; also, if you're using a CD-RW disc, make sure that it's completely formatted.)

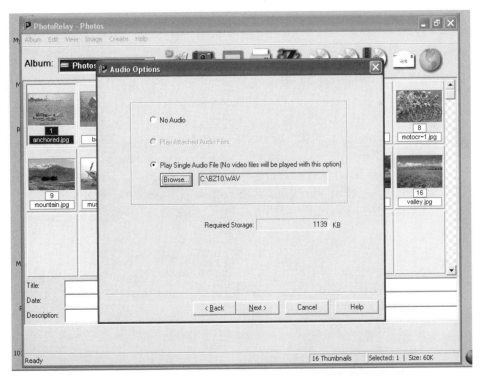

FIGURE 11.14 Adding a background audio track to your slide show.

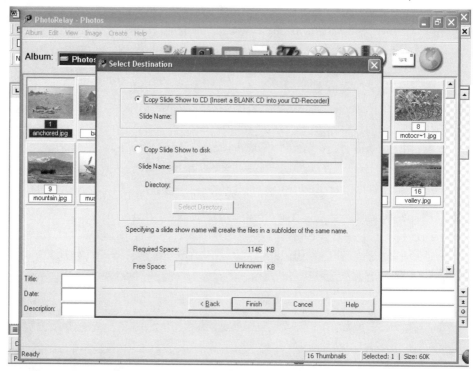

FIGURE 11.15 Will that be disc or hard drive?

tip To test the slide show before you record it, click Copy Slide Show to disk. Specify a Slide Name and click Select Directory to choose the location for the files—this way, you can run the show from your hard drive to check things out before using a disc. (Note, however, that you'll have to follow this entire project procedure again to record the slide show on CD.)

7. To create the slide show in a separate folder, click the Slide Name field and enter the name you want; however, I recommend that you leave the Slide Name field blank and create the slide show in the root (top level) of the disc; this way, others won't have to change to the folder manually before running the show.

8. Click Finish to start the recording—PhotoRelay will automatically load Easy CD Creator to handle the actual burn!

Once the recording has finished, it's easy to run your slide show: Load the disc into a computer CD-ROM drive. If you chose to record the files in the root directory of the CD, Windows will automatically recognize the disc and start the show!

 tip **To run the show manually, open the CD, locate the file SLIDE.EXE and double-click it.**

Project: **Creating a Video Postcard on CD**

Our second PhotoRelay project concerns digital video—along with the CD-based screen show you just produced, this surprising program can also create a professional-looking *video postcard* that you can send to others on a CD-R or CD-RW disc. (This is a neat way to share digital video with those of your family and friends who don't have a DVD player.) In fact, you can even capture the digital video from any device that's compatible with the Video for Windows AVI video format! The video postcard can be shown on any PC running Windows 95 through Windows XP.

Requirements

- A digital video source (or a digital video clip that's already been saved to your hard drive)
- A blank CD or CD-RW

Because I've already discussed digital video transfer and capture earlier in the book, we'll assume that you're using a digital video clip you've already saved to your hard drive.

✔ **Follow these steps to create the postcard:**

1. Run PhotoRelay.

2. Click the Make Postcard icon on the toolbar (which looks like a CD-ROM, overlaid with a strip of movie film). PhotoRelay displays the Select Template dialog box you see in Figure 11.16. Choose the category for the postcard background and click one of the background thumbnails to select it—alternatively, you can click on From Image and click the Select Image to import your own picture into PhotoRelay as the background for the postcard.

3. Choose the size of the video clip—PhotoRelay adjusts the "window" within the background to match the dimensions you choose. As you might expect, a larger window of 320 × 240 results in a much better postcard, but the larger video clip may not run smoothly for older PCs. Click Next to continue.

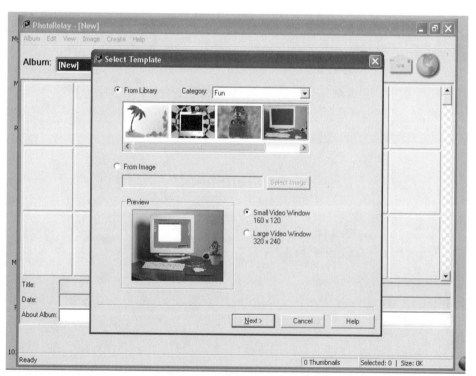

FIGURE 11.16 Choosing the background and clip size for our postcard.

4. PhotoRelay displays the Postcard Message dialog you see in Figure 11.17. Click Select Font to choose the font for your text, then type your message in the left side of the "virtual postcard." You can also enter your name and the name of the recipient by clicking in the To: and From: fields on the right side of the card. Click Next to continue.

5. Time to select the video clip: Click the Select Video button (Figure 11.18), navigate to the location, click on the clip, and click Open. (The program accepts digital video in .AVI, .MOV, and .MPG formats.) You'll notice that PhotoRelay displays the familiar VCR Play and Pause buttons next to the video clip filename; click on Play to preview how your clip will look.

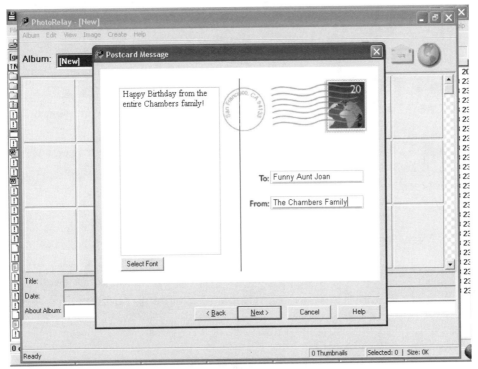

FIGURE 11.17 I guess a "virtual postcard" needs only 20 cents postage!

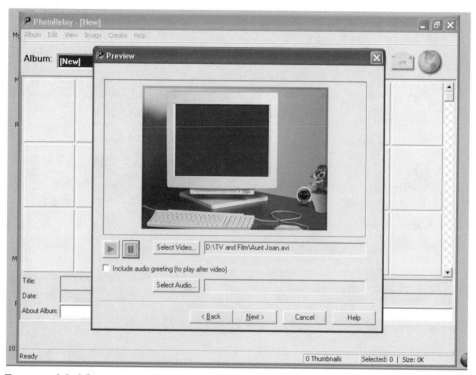

FIGURE 11.18 Selecting the digital video and audio clips to use in your postcard.

tip **If you're using another video format that's not recognized by PhotoRelay, don't despair—that video clip can usually be converted into one of the supported formats, using a video editing program such as Adobe Premiere or ArcSoft's ShowBiz.**

6. Why not personalize your video postcard with your voice, as well? (You can say hello to Grandma yourself!) If you'd like your postcard to play a digital audio clip after the video is done, click the Include audio greeting checkbox to enable it, and click the Select Audio button to choose an audio file in .WAV or .MP3 format. Click Next to continue.

7. The final PhotoRelay wizard dialog displays the Select Destination dialog shown in Figure 11.19. Load a blank CD-R or CD-RW disc into your recorder. (As with the slide show CD you created earlier, do not use a disc with an existing session; also, if you're using a CD-RW disc, make sure that it's completely formatted.)

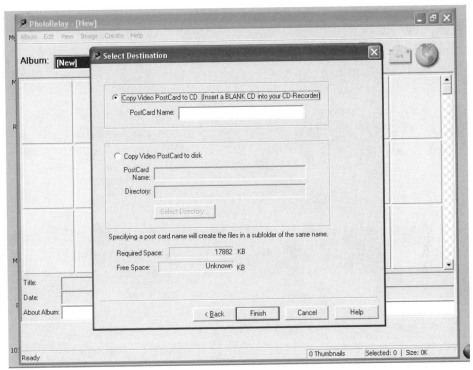

FIGURE 11.19 Setting options before burning the video postcard.

tip Click Copy Slide Show to disk to try out the postcard
before you record it. Click PostCard Name to choose a folder
name for the files—PhotoRelay will create the postcard on your hard
drive, so you can run it without recording a disc. (You'll have to
follow this entire project procedure again in order to burn the
postcard on a CD.)

8. It's a good idea to leave the PostCard Name field blank and record
    the postcard in the root (top level) of the disc; this way, others
    won't have to manually change to the folder, and it should start
    automatically under Windows. However, if you want to place the
    postcard files in a separate folder, click the PostCard Name field
    and enter the desired name.

9. Click Finish to start the recording, and watch as PhotoRelay
    launches Easy CD Creator to record the disc.

It couldn't be easier to run the postcard: Simply load the disc into a computer's CD-ROM drive. If you chose to record the files in the root directory of the CD, Windows should automatically recognize the disc and display the card!

 tip **To run the show manually, open the CD, locate the file VPLAYER.EXE in the root directory, and double-click it.**

Summary

This chapter focused on the multimedia program PhotoRelay and how you can use it to create spectacular self-running slide show CDs and video postcard CDs from your digital pictures and video.

12

Making Movies with ArcSoft ShowBiz

In This Chapter

✔ Opening and creating albums

✔ Adding elements to the storyboard

✔ Adding transitions

✔ Using the timeline

✔ Applying effects

✔ Adding text

✔ Previewing your movie

✔ Saving your movie to disk

In Chapter 6, "Recording a DVD with Existing Files," I discussed how you can create your own DVD video discs and cDVD discs using MyDVD; we used digital video files that you had already created and stored on your hard drive. Often, these video files are simply raw DV footage that you've copied directly to your computer from your DV camera or video that you've captured to your hard drive from an analog source.

But what if you want to create your own cinematic masterpieces? What if you want to edit that footage and add things such as special effects, still images, or even your own separate audio track? Now you're talking about a video editor: a program that's specifically designed to help you take the various "building blocks" of a movie and combine them to create an entirely new and original movie.

In this chapter, I'll show you how to use ArcSoft ShowBiz, the easy-to-use video editor that's bundled with many HP DVD-Writer drives. You'll discover how professionals use a storyboard and a timeline to "assemble" a movie, and you'll build your own original wedding video!

Introducing ShowBiz

The ShowBiz main window that you see in Figure 12.1 has four main parts:

1. **Player window.** This window can display your movie at any time while you're editing; it can be resized to several different dimensions.

2. **Media library.** The tabbed list at the upper right of the ShowBiz window allows you to choose elements from four different media libraries that you can add to your movie: video, still, and audio media; transitions between scenes; special effects; and text. You can also display miniature thumbnails of the items in your library: Click the Album view mode button at the upper right of the library list to toggle the display of the thumbnails on and off.

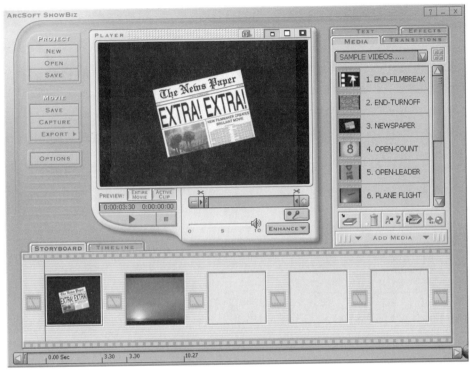

FIGURE 12.1 The ShowBiz main window.

3. **Storyboard strip.** This strip at the bottom of the window allows you to add elements to your movie from the media library.

4. **Timeline strip.** The timeline display usually hides behind the storyboard strip—click the Timeline tab, and you can alter the length of effects and transitions or edit your movie's audio track.

Although ShowBiz doesn't use a traditional menu system, common command buttons for opening and saving projects and movies are grouped at the upper left of the window. To display the program's help system, you can press F1 at any point.

tip **You can also learn more about ShowBiz by using the Showbiz Basics help system, which you can open by clicking on the child's "building block" that appears on the main window. You'll find that the Basics dialog will answer most of the common questions that a novice will ask about the program.**

Expanding Your Media Library

Before I get to the actual movie creation process, you should learn how to add your own video and still elements to the ShowBiz media library—otherwise, you'll be creating great movies, but you'll be limited to the sample elements that ArcSoft provides! (As you'd imagine, this can get pretty boring pretty quickly.)

tip **ShowBiz Media elements are stored in *albums*, which are actually collections of links to multimedia files on your hard drive—therefore, it's important that you don't add a video clip or still image to your media library until you've stored it where you want it on your system! To illustrate, if you add a digital video clip from your \My Documents\My Videos folder to your ShowBiz library, then move the clip to another folder on your hard drive, you won't be able to access the clip in ShowBiz; instead, it shows up with a red "unavailable" flag within the media library.**

✔ **To add your own video or still image elements from your hard drive, follow these steps:**

1. Click the Media tab within the media library, and click the Add button that appears under the list (it looks like a book with an arrow).

2. ShowBiz displays the standard Windows Open file dialog. Navigate to the location of the file you want to add, click on the filename to select it, and click Open.

3. The program displays the dialog you see in Figure 12.2, prompting you either to create a new album or to select an existing album that you've created. (You can't add media elements to the Sample albums.) To create a new album, click <New Album>; to select an existing album, click on the drop-down list, and select the destination.

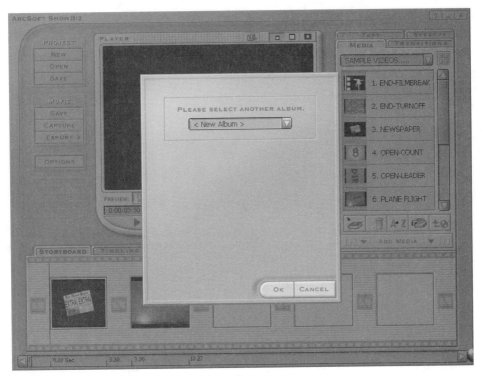

FIGURE 12.2   Creating a new media album.

4. Depending on the format of the element you've added, ShowBiz may prompt you for scene or image options; leave these options set to their default values, and click OK.

5. The new album (or the existing album) with the new element appears in the media library. If you've created a new album, you can rename it by clicking in the album name at the top of the list and typing a new name.

tip You can also download new media library elements from ArcSoft's Web site; click the Download new content button at the bottom of the media list (it looks like a globe with an arrow). Windows will open your Web browser, and you can download additional media files to your PC.

To acquire a still image from a TWAIN-compatible scanner or digital camera, click the Acquire button, and select the image source. You can also sort the elements in an album by clicking the Sort button, which displays the different sort criteria you see in Figure 12.3. Click the sorting option you want and click OK.

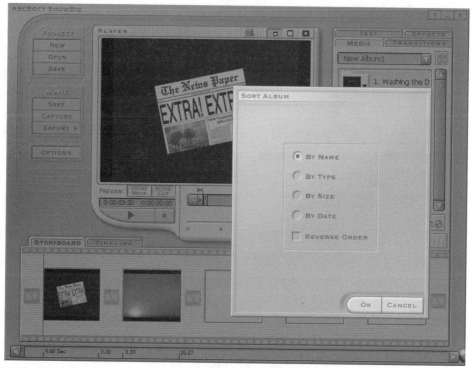

FIGURE 12.3 You can sort album elements by four different criteria.

Building a Movie from Media Elements

Creating a movie is easy within ShowBiz: You add the elements that you want to appear in linear order, moving from left to right. Usually, these media elements will be video clips and still images that you've added to your library, but they can also include text titles, *effects* (where a special look is applied to an image or a scene) and *transitions* (an animation or special effect that occurs between scenes and images).

✔ **To create a movie using the storyboard strip, follow these steps:**

1. If you've been working with ShowBiz on another project, click New under the Project menu—ShowBiz gives you the option of saving the existing project before closing it.

2. Decide what you want to appear first in your movie—either a video clip or a still image—and click on the Media tab in the media library to display the elements of that type. To switch to another album, click the drop-down list arrow next to the album name.

> *tip* **To view a media element before you decide to add it to the storyboard strip, right-click on the element in the list and choose Preview Media from the pop-up menu. ShowBiz displays the element in the Player window.**

3. ShowBiz offers two methods of adding the element: You can either click and drag the element from the media library to the storyboard, or you can click the element you want to add and click the Add button at the bottom of the media library display. Either way, it appears in the next open media square on the storyboard. Figure 12.4 illustrates a still image I've added as the first element in the storyboard display.

Once you've added a number of still images and video clips to your storyboard, you can click and drag elements from one media square to another to change their positions (great for those times when you discover that you've put your aunt on-screen before your mother). To delete an element from the storyboard, right-click the offending element and choose Delete. Note that you can also choose to delete all video clips and still images at once, in case you want to start over from scratch.

> *tip* **Need information about a clip or image that you've already added to the storyboard? Right-click on the element in the storyboard and choose Properties from the pop-up menu, and ShowBiz will display the relevant identifying information for that element.**

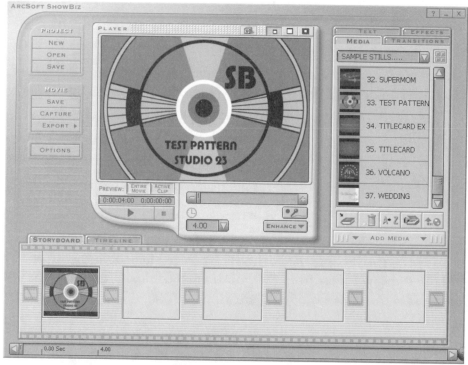

FIGURE 12.4   *I'll use a still image to begin my movie.*

Adding Transitions

No matter how visually interesting, any film can be enhanced by the use of animated, moving transitions between elements—personally, my favorite transitions are the old "twirling rainbow Bat symbols" you probably remember from the Batman TV series (and yes, you can set those up with a Spiral transition). Once you've added at least one still image or video clip to your storyboard, you can place a transition.

Note, however, that it's easy to add too many transitions to your project! Use restraint, and your viewers will appreciate it. Watch for transitions in your favorite TV shows and commercials, and you'll see that there's certainly no need for a transition at every scene change.

✔ **Follow these steps:**

1. Click the Transitions tab in the media library to display the list of transitions—remember, you can click on the album drop-down list to choose from different types of transitions. When you rest your mouse pointer on top of a transition for a second or two in the library list, the example image animates to show you how the transition will appear.

2. Click the entry in the media list for the transition you want to use.

3. Click Add Transition to copy the transition into the next open transition square on the storyboard; transition squares are smaller than media squares, and they carry a filmstrip icon with a diagonal cut.

tip **If you're pressed for time, but you'd still like to add a different transition between each element in a movie, right-click any media or transition square on the storyboard strip and choose Random Transition to All; this adds a transition that ShowBiz chooses at random between each video clip and still image. To duplicate the same transition throughout the entire movie, right-click the transition square that contains the transition you want to replicate, and choose Apply Transition to All.**

Editing with the Timeline

Once you've added the video clips, still images, and transitions to your film, you might think you're done and ready to review your work—but ShowBiz offers you the ability to add creative touches with the timeline strip display. The storyboard strip is a linear representation of the elements that make up your movie—the timeline strip allows you to add:

- Special effects
- Animated text
- Background audio

You can also use the timeline to adjust the starting and ending points for these additions. To display the timeline, click the Timeline tab at the top of the strip; Figure 12.5 illustrates the timeline for a movie I'm working on. Note that you can still see the media elements from the storyboard; however, the transitions may be a little harder to pick out, because they are actually combined with the media elements in the timeline view. If you get confused and forget what's been added on the storyboard side, place your mouse pointer over a media element for a second or two, and ShowBiz will remind you of any transitions that you've added.

Note that four additional rows appear on the timeline strip: above the strip are the text and effects displays, and below you'll find two audio track displays. Again, these are linear controls—they help you "block" the location and duration of each type of edit you can make.

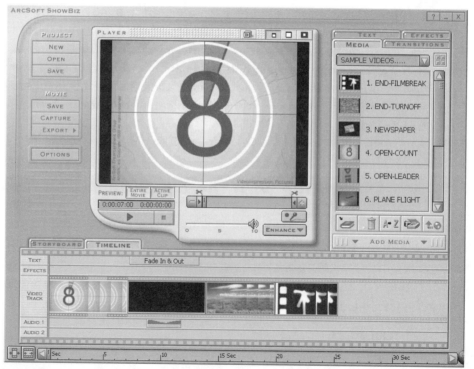

FIGURE 12.5 Preparing to add effects, text, and audio using the timeline strip.

tip When you're editing, remember that experts in the movie industry use only about 10% of the content they shoot! If you leave that camera running because you don't want to miss that special moment, you'll end up with a tape full of noncritical footage that you and your friends and family may not want to see. Keep your audience's attention span in mind while you're editing your video clips.

Creating a Different Look with Effects

Effects can add an entirely new dimension to all of your media elements—for example, one of my favorite effects, Film Grain, is perfect for turning that video clip you shot yesterday into "antique" 8-mm movie footage! Other filters you'll use often include vertical and horizontal hold "problems" (in the TV album), wacky frames that you can add around the borders of a clip (in the Fun Frames album), and the Blur and Ripple effects in the Filters album.

✔ **Follow these steps to add an effect to your movie:**

1. Add all of the stills and video clips you need for your movie, then click the Effects tab in the media library to display the list of effects. To switch effects albums, click on the album drop-down list box.

2. Click the entry in the media list for the effect you want to use.

3. Click Add Effect to copy the effect into the next open block on the Effects row—remember, the Effects row appears alongside the timeline strip; therefore, ShowBiz automatically switches to the timeline strip display (if necessary) as soon as you add an effect or a text title.

4. Drag the beginning and ending edges of the effect to specify where the desired effect will start and when it will finish during the movie—note that you can also click and drag the entire effect

block along the row to wherever you like during the movie. You can have only one active effect at a time, but the effects can follow each other in sequence.

Adding Text to Your Movie

There are two ways to add a text title to your movies:

1. **Using a still image.** You can use an image editing program such as Photoshop or Paint Shop Pro to create a still image that has the title you want—you can then add the image to your media library, and add it to the proper media square so that it appears at the right time.

2. **Adding a text element.** ShowBiz includes a number of preset animated text effects that you can use to display your message during your movie.

As you can imagine, the Text elements inside ShowBiz are much easier to add than a still image—you don't have to open up another application or add a special image to your media library—and they're much more interesting from a visual standpoint.

✔ **Follow these steps to add text to your movie:**

1. Once you've finished adding stills and video clips to the storyboard strip, click the Text tab in the media library. Click on the album drop-down list box to select the type of text effects you want to use.

2. Click the entry in the media list for the type of text you want to add.

3. Click Add Text to copy the effect into the next open block on the Text row, which appears at the top of the timeline strip. ShowBiz opens the Text panel you see in Figure 12.6, where you can type the desired text, specify the number of seconds it remains on-screen, and set options such as the font, alignment, type size, color, and shadowing.

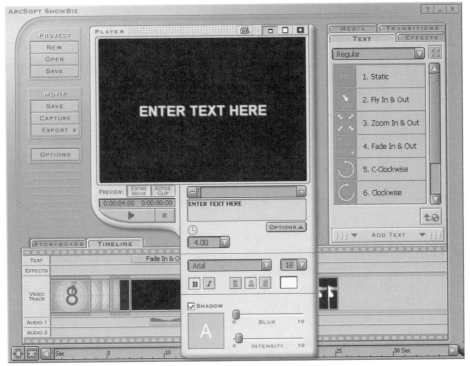

FIGURE 12.6 Specifying the text options for text that will fade in and out.

4. Once you've typed the text for the title and set the type options, click the Options button to return to the timeline.

5. As you did with the effects row, you can click and drag the beginning and ending edges of the text block to specify where it will appear.

Audio Tracks

If you like, you can add one or two audio tracks for your movie—for example, one track might include your narration of events in your video, and the other track could be background music. Note, however, that it isn't necessary to add audio (your video clips will likely carry their own audio, and the media clips and animation provided with ShowBiz also carry their own audio). You can insert audio tracks from the Media tab (choose the Sample Audio album)—both .MP3 and .WAV files can be added to your library.

In fact, there's a trick you can use to add more than two audio tracks to your movie: Just save the file and reopen it! For example, let's say you have three audio tracks you want to use. Add the first two audio tracks in the Showbiz timeline, then save the file as an MPEG2 video clip. Now you can add your new clip (which already contains the first two tracks) into your ShowBiz media library, and you can add your third audio track! (Note that when you save the file to a MPEG2 file or other video format, you can no longer change the edits you made previously.)

Previewing Your Work in ShowBiz

When you're ready to review your work by previewing your movie, ShowBiz can preview the project in the Player window in two ways:

1. **View a single clip.** Click a media square on the storyboard strip and click the Active Clip button (under the Player window) to view a still image, video clip, or transition. You can also use this setting to view effects and text or listen to audio tracks while you're using the timeline strip.

2. **View the entire film.** Click the Entire Movie button under the Player window to view the entire movie—typically, this will display the "Processing" progress bar that you see in Figure 12.7, which means that ShowBiz must *render* (or "build") the effects, transitions, and media elements into a single video stream. This process usually occurs each time you view the entire film if you've changed or edited anything in your movie: The longer the movie and the more effects and transitions you use, the longer the processing will take.

tip Make sure that you hit the Stop button after you view the entire movie to ensure that the cursor is placed at the beginning of the timeline.

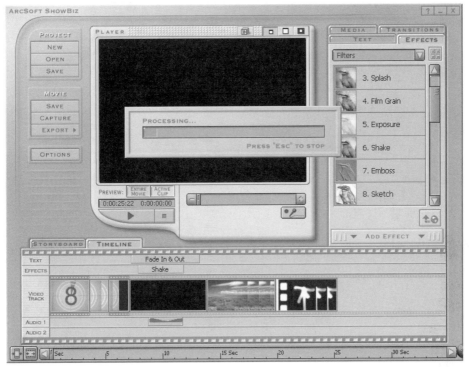

FIGURE 12.7 ShowBiz must process effects and transitions before you can watch your entire movie.

Once you've selected the proper preview mode (and any process-ing has completed), you're ready to watch: Click the green Play button underneath the Player window. Note that ShowBiz displays a moving vertical yellow line to indicate the current point in the storyboard or film strip; this can help you "debug" spots where an effect fades too fast or an audio clip doesn't play at the right time. Click the yellow square Stop button to stop the action at any time. To jump forward or backward, click and drag the moving slider underneath the Player win-dow in the desired direction.

You can expand the ShowBiz Player window using the three but-tons at the top right corner of the ShowBiz window; the default is the Normal view, but you can also switch to Large view (shown in Figure 12.8) and a full-screen display.

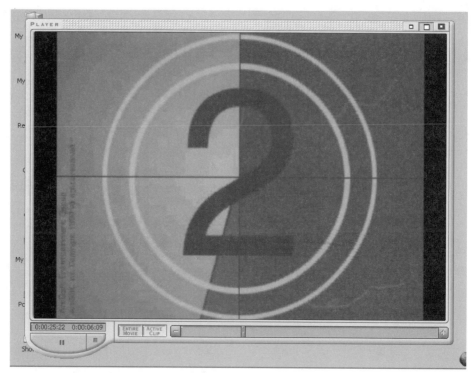

FIGURE 12.8   Expanding the Player window to a larger size helps when you're reviewing your project.

The first time counter above the Play and Stop buttons displays the total duration of the clip or movie, and the second displays the elapsed time.

Saving Your Classic to Disc

Ready to immortalize your cinematic work? ShowBiz gives you a wide range of possibilities for creating your final movie—note that this is different from simply saving the project to your hard drive (which you can do from the Project group at the top left corner of the ShowBiz window).

Final movie options include:

• **Saving your movie as a digital video file.** Click Save in the Movie group at the upper left of the ShowBiz window to display the dialog you see in Figure 12.9: From here, you can save your movie in MPEG1, MPEG2, QuickTime MOV, and Windows AVI formats. You can even save your movie as raw DV footage or in an executable format that you can send as an e-mail attachment. Type a filename and click Browse to select a location where the file will be stored, then click the Format drop-down list box to choose the format—click Settings to specify options such as the frame size and audio quality. When you're ready to save, click OK. If you're creating a DVD or cDVD, use MPEG2—for video that's destined for use in a VCD, choose MPEG1. It's important to remember that switching compression standards will cost you image quality and detail in your movie, so choose the right codec from the start (and always save your project so that you can open it and export it again in ShowBiz if necessary)!

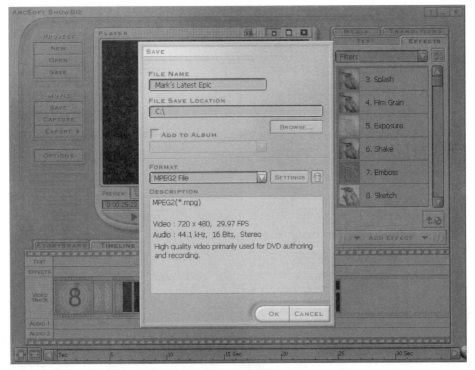

FIGURE 12.9 Saving your movie as an MPEG2 file.

tip Need to send your movies overseas to someone using the PAL video standard? No problem: ShowBiz can also convert NTSC video to PAL or PAL to NTSC! Click the Options button and choose the desired standard when saving or exporting video.

tip To add your completed movie to the Media Library as a video clip for future projects, click the Add to Album checkbox to enable it, and click the Album drop-down list box to specify the album where the new clip will appear. (This is a great way to create a standard "title" sequence that you can use at the beginning of all your movies.)

- **Capturing your movie to other video hardware.** If your PC has video capture hardware or a FireWire connection, you can send the signal to an analog VCR, analog camcorder, or DV camcorder and record it. Click Capture in the Movie group to display the Capture dialog you see in Figure 12.10. Click the Video and Audio Device drop-down list boxes to select the output hardware, and use the Play, Record, and Stop buttons to control the playback of your ShowBiz movie. (The exact procedure you follow will depend on the type of capture connection you're using; refer to your capture hardware documentation and the ShowBiz Help system for specifics.)

- **Exporting your movie to CD, DVD, DV tape, or MyDVD.** ShowBiz makes use of MyDVD, the application that I cover in detail in Chapters 6 ("Recording a DVD with Existing Files") and 7 ("Direct-to-DVD Recording"), to copy your movie directly to several different types of digital media. Move your mouse pointer over the Export button in the Movie group and click on the desired output—you can even create a MyDVD project from your completed ShowBiz movie! Depending on the type of exporting you chose, ShowBiz will prompt you for the required settings; follow the on-screen instructions.

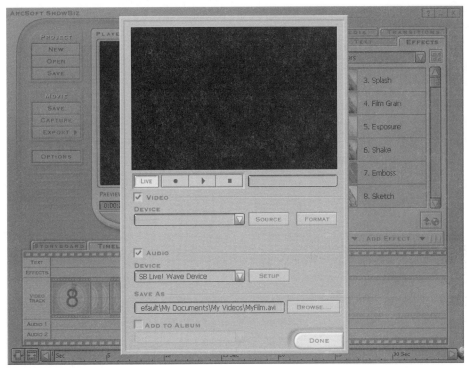

FIGURE 12.10 Capturing a ShowBiz movie to digital or analog video hardware.

Summary

In this chapter, I discussed ArcSoft's ShowBiz, and you learned the basics of movie editing in the digital world. We covered how to create your own movie using your digital video clips, still images, transitions, effects, and animated text, and I showed you how to save your movie to hard drive or import it to MyDVD.

Troubleshooting Recorder Problems

In This Chapter

✓ Solving common hardware problems

✓ Solving common software problems

✓ Solving buffer underrun problems

The goal of this chapter is a simple one: to help you diagnose and solve both common hardware and software problems you may encounter with your recorder. Techno-types call this process troubleshooting—it's usually frustrating and often complex, but the good news is that you can successfully troubleshoot virtually all of the possible tribulations that are likely to crop up with your recorder.

Whenever possible, I've provided cross-references to relevant chapters throughout the rest of the book.

Common Hardware Problems

In Chapter 2, "Installing Your DVD Recorder," I covered possible installation problems you might encounter. In this section, I'll discuss various hardware-related problems that can crop up with any type of recorder.

The recorder keeps telling me it's not ready (or it won't eject the disc)

This problem can be particularly irritating—and, if the disc doesn't eject at all, not just a little bit frightening, as well. Check these solutions:

- **Mounting delay.** Some recorders take longer than others to recognize a blank disc (or, if you're trying to read an existing audio CD, to extract tracks). Wait until the drive is ready before continuing—it's a good idea to watch the front panel light to see whether it's flashing (your recorder's documentation will have the specifics). Although it doesn't happen often, it's possible to load a disc off-center in the tray, which can cause circular scratches on the surface of the disc—make sure that you take a second when loading a CD to verify that it's setting in the center of the drive tray!

- **Lead-in/lead-out recording.** Your drive may still be recording the lead-in or lead-out portions of the disc; these delays occur at the beginning and at the end of a CD recording session. Unfortunately, many recording programs don't indicate that they're writing these areas, so it's easy to think that your computer is locked up. Give your recorder at least a couple of minutes to finish this processing.

- **Check your SCSI termination and device IDs.** If you're using a SCSI recorder and you've recently changed your SCSI settings or device chain, there's likely a problem with your computer's SCSI hardware. Run the diagnostic software supplied by your SCSI adapter card or drive's manufacturer to verify that you've correctly installed the drive.

- **Reinstall your recorder's drivers.** Have you recently been experiencing lockups, or has your computer been displaying error messages that you haven't seen before? Your drive's software drivers may have been corrupted or deleted because of a power failure or a misbehaving program. Reinstall the software drivers that you received with your drive, or download the updated drivers from the manufacturer's Web site and install them.

I can't read the discs I record on one particular CD-ROM/DVD-ROM drive

On the positive side, this particular problem indicates that there's nothing wrong with the discs you're recording—instead, look to the drive you're using to try to read the disc as the source of the trouble. Consider these possible culprits:

- **Drive is not a MultiRead model.** CD-ROM drives built before the days of the MultiRead specification can't read a CD-RW disc, so record the data on a standard CD-R, instead. DVD-ROM drives don't have this problem with recorded CDs, but don't forget that a DVD+RW drive may not be able to read a DVD-RW disc—the reverse is often true, as well.

- **Media incompatibilities.** You may have noticed that some CD-R discs use different combinations of dye color and reflective layers; recently, for instance, I've used discs with blue, gold, purple, green, and even pink dyes! Technically, this is *not* supposed to be a problem (any drive should be able to read these discs), but I have personally encountered drives that basically "don't like" a certain dye color. For example, a drive may be able to read a recorded CD that uses a blue dye layer and silver reflective layer, but it may not reliably read a recorded CD with a green dye layer and a gold reflective layer. Try reading a disc recorded with another combination of dye color and reflective layer.

- **Alignment problems.** It's possible that the drive is incorrectly calibrated; if this is the case, it will probably have problems reading *any* discs you've recorded. If the drive returns errors with a commercially manufactured disc of the proper type—for example, if a CD-ROM drive can't read any audio CDs or a DVD-ROM drive can't read any DVD movies that you've bought, bring the drive to your local computer repair shop and ask the technician to check it.

Common Software Problems

Which is worse—hardware problems or software problems? That's a common question I hear all the time, and the answer is both! Windows, Mac OS, and Linux are difficult beasts to tame if a driver is corrupted or a software installation overwrites your recording application.

In this section, I'll describe a number of common problems due to software and how you can solve them.

My computer locks up when I try to record

Unfortunately, this error can be due to problems outside your recording application and its associated files—if Windows has become unstable because of some other problem, for example, the source of the trouble isn't your recording software. Try these tricks and solutions:

- **Reserve sufficient free space on your hard drive.** Remember, both Windows and your recording software need additional space for temporary files to record a disc successfully. I recommend that you reserve at least 200 or 300 MB of free space to avoid recording errors and lockups.
- **Read your log file.** Because your computer locks up, you're not likely to see an error message; this makes the job of troubleshooting even more difficult. Use the log file option if your recording software supports it—the log file may contain error messages that will help you in troubleshooting the problem.

- **Reinstall CD recorder drivers.** If your recorder doesn't work at all, your drive's software drivers have probably been corrupted or deleted. Reinstall the software drivers that you received with your drive or download the updated drivers from the manufacturer's Web site and install them.

My recording software reports that a blank disc isn't empty

This same problem may cause your computer to report that the disc doesn't have enough room or is incompatible with the format you're trying to use. Check these solutions:

- **Wrong disc type or overburning.** Older CD recorders aren't compatible with CD-RW discs, but these drives are museum-quality antiques in 2002! These days, this error message can be caused by attempting to overburn 80-minute "extended" discs in some recorders.
- **Previous session.** Check to make sure that you're not running out of room on the disc because of a previous session that you've recorded. Within Easy CD Creator, you can click CD and select the CD Information menu item to display the disc summary, which will show if that disc is really blank.
- **Bad media.** It's possible that the blank disc you're trying to use is bad; try recording with another disc from another manufacturer.
- **Reformat.** If you're writing a DVD+RW disc, DVD-RAM, or CD-RW disc, this error message often indicates either that the disc has not been formatted or that a previous formatting operation was interrupted before it could complete. Reformat your disc to refresh it.

I get a fixation error when I try to write a new session to a CD-R disc

In effect, your recording software is complaining that it can't close the last session on the disc. You may receive this message if the recording session is interrupted by a power failure or your computer locks up. Unfortunately, the disc may be unreadable; however, you can try using

Session Selector to read data from the disc. (If it's a multisession disc, you should be able to read data from previous sessions this way.)

tip **If you're using Easy CD Creator, you may be able to recover some or all of the data, even if the fixation process didn't complete properly. Click CD and select the CD Information menu item to display the disc summary. If your recorder is able to recognize the data on the disc, use the Recover feature to attempt fixation again.**

I can't record MPEG, AVI, MOV, or ASF clips to a DVD or Video CD

These problems occur while you're using MyDVD (Chapter 6, "Recording a DVD with Existing Files") or VCD Creator (Chapter 10, "Recording Advanced Formats and Video CD Discs"). Possible solutions include:

- **Compression/codec problems.** Your recording software may not support the compression and codec used to record a particular video clip. If possible, use a video editing program to convert them to the standards supported by your recording software.
  - If you suspect that a video file you imported is using a codec you don't have, try to play it in the built-in Windows Media Player. If the file plays, then you have the codec in your system; otherwise, the program will automatically attempt to download the required codec. This often works, but it won't help if the video clip was recorded using a hard-to-find commercial codec.
- **Unsupported formats.** Both MyDVD and VCD Creator recognize AVI and MPEG, but there are numerous other, more exotic video formats that these programs can't handle. Again, use a video editing program to convert them, if possible, or use the application that created the clip to save the files in MPEG or AVI format.

Windows (or Mac OS) long filenames on my hard drive are truncated on discs I record

- **Long filenames not enabled.** If you're recording a disc for a Windows system, use Microsoft Joliet file format, which enables long filenames, instead of ISO 9660 file format. If you're writing a Mac disc, check the documentation for your recording software to determine where you can enable support within the program for long filenames. (Note that all of the software that ships with HP DVD-Writer drives defaults to long filenames.)

My recording software says I can't create a nested folder

This one's pretty easy: Use Microsoft Joliet (under Windows) or Apple Extended (under Mac OS). You're attempting to create a disc using the ISO 9660 file format, which won't allow more than eight levels of nested folders. Use Microsoft Joliet, if possible, or disable DOS/ISO 9660 conventions. You also have a limitation of 255 characters on your pathname under ISO 9660.

My DVD-ROM or CD-ROM drive doesn't recognize the disc I just recorded

There are several possible solutions to this problem—note that some of these problems apply only to CD recording.

- **Session remains open.** If you recorded a session on a CD-R or CD-RW and left it open, a standard CD-ROM drive will not be able to read it—you either have to close the session using Easy CD Creator or read the disc in your recorder. Programs such as MyDVD will close the disc automatically when writing to CD or DVD.
- **Unrecognized format.** Depending on the recording format you used, a disc created on the Macintosh may not be recognized by a PC. Consider recording cross-platform discs in true ISO 9660 format.
- **Dirty disc.** Dust, grease, and oil on the surface of a DVD or CD can prevent your drive from reading data—the section titled

"Cleaning Discs the Right Way" in Chapter 1 shows you how to care for discs.

- **Auto insert notification disabled.** If a program is supposed to run automatically when you load the disc, try using Windows Explorer to run the program from the disc—if you can, then Auto insert notification has been disabled within the Properties panel for the drive you're using.

- **Faulty recording.** You can use the verify feature within HP RecordNow or Easy CD Creator to make sure that the disc is readable (although rare, it is possible for a recording program to indicate that a disc has been recorded successfully when, in fact, something went wrong during the process).

I hear a click between every track of the audio CDs I record

This is a very common problem when recording audio CDs using MP3 files, but it usually appears only on some audio CD players—in fact, you may be able to listen to the disc on another player without any problem at all. You should be able to eliminate these clicks by recording your audio CDs using Disc-at-Once mode, which enables the laser to record the entire disc without shutting off between tracks—this area between the tracks is causing the clicks. If you're using the HP Record-Now program, digital audio is always written as DAO.

An MP3 track I recorded plays in mono instead of stereo (or it's "chopped off")

This problem isn't due to your recording hardware or software at all! Instead, you used a substandard MP3 file. Not every MP3 file you receive over services like Napster or LimeWire is recorded with CD quality—in fact, the original audio may have been recorded in mono. Also, it's a good idea to preview an entire MP3 track before you record it, because these files are often badly edited!

My Linux/UNIX system can't read the discs I've recorded

Again, beware of the cross-platform compatibility issue with Linux and UNIX computers. These tips should help:

- **Use ISO 9660.** Most older variants of these operating systems don't recognize Joliet or Mac HPFS file formats, so use ISO 9660 file format when recording for Linux/UNIX.
- **Watch your case-sensitivity.** Linux and UNIX are case-sensitive, which may cause problems when running programs or accessing data files directly from the disc.

I can play the audio tracks on a CD-Extra disc, but my computer can't read the data track

This problem turns a perfectly good CD-Extra disc into a plain audio CD. Check these potential fixes:

- **Unsupported format.** Your computer's older CD-ROM drive probably doesn't support the CD-Extra format—see whether you can read the data tracks on a late-model CD-ROM.
- **Disc incorrectly closed.** The data track may have been incorrectly saved. If you're using Easy CD Creator, you can click CD and select the CD Information menu item to display the disc summary, which will show whether that disc is really blank. If your recorder is able to recognize the data on the disc, use the Recover feature to attempt fixation again.

I can't read Kodak PhotoCDs on my computer

No secret on this one: Discs created for use on a Kodak PhotoCD player can't be accessed on a computer, and your recorder can't create Kodak PhotoCDs—they use a proprietary format.

Eliminating Buffer Underrun Errors

If your recorder doesn't support burn-proof recording, even the fastest computer may fall victim to the dreaded buffer underrun error; as you learned in earlier chapters, this error results when your computer can't transfer data to the recorder at the necessary rate to ensure a successful recording. The older and slower your computer, the more likely it is that a disc will be ruined. Try these solutions:

- **Record from a disc image.** Rather than recording a number of smaller files "on-the-fly" (which is far less efficient), try creating a disc image, then record from the image—this is much more efficient and requires less overhead, because the image is a single file.

- **Record at a slower speed.** Although your system may not be able to support the highest speeds available with your recorder, you can always record at a slower speed—for example, if you can't record at 32x, try recording at 16x, instead.

- **Don't multitask while recording.** Dedicate the entire resources of your computer to recording by shutting down all unnecessary applications (including those running on the Windows taskbar).

- **Defragment before recording.** A defragmented drive transfers data significantly faster; I describe how to defragment a hard drive running Windows XP in the section titled "Defragmenting Your Hard Drive" in Chapter 3.

- **Use the fastest drive on your system.** Whenever possible, choose the fastest drive on your system as the location for source data and temporary files created by your recording software.

tip Even if you use all of these tricks, an older computer may still not be able to handle the data transfer speeds required by today's DVD recorders. Don't give up hope, however; you can still try to record that material! There is a way to make sure you don't lose a blank CD-R or DVD-R disc: Use your recording software's simulation or test mode first, before the actual recording begins. If the data can be successfully recorded in test mode, you can be certain that there will be little likelihood of errors during the recording process. For example, MyDVD allows you to record the image to the HD instead of disc. If that works, all the encoding has been done, so it will take less processing time (and will significantly lower the chance of an error) when you burn the image from the hard drive to the disc.

Summary

This chapter covered the common hardware and software problems that often occur while recording CDs and DVDs, as well as recommendations on how to troubleshoot and avoid them.

14

Adding an HTML Menu System

In This Chapter

✔ Determining whether your disc needs a menu

✔ Adding menus to a DVD

✔ Introducing HTML as a menu language

✔ Designing a menu system

✔ Comparing HTML creation software

✔ Project: Building an HTML Menu

In Chapter 9, you learned how to add labels and jewel case inserts to enhance the appearance of your discs—with a good labeling program such as CD Label Creator, you can produce a professional-looking CD-ROM that will impress your friends, co-workers, and customers.

To be honest, however, that's just the outside of your project! As the old adage goes, "You can't tell a book by its cover"—even the nicest-looking project disc can suffer from bad organization and a lack of those user-friendly convenience features that are common among software that's commercially produced. For example, if you're distributing MP3 music files, you might want a button on your disc that installs those music files to the computer's hard drive.

You can help prevent "data chaos" by using folders and the common-sense guidelines I covered in Chapter 3 (in the section titled "Organizing Files"), but what if the person who'll be using your disc has no computer experience at all? If you're a software developer, you'll probably design and write a custom menu system using Visual Basic, C, or REALbasic that will make the disc easy to use and navigate...but what if you have no programming skills? You may end up shaking your head and giving up entirely on adding any menu system to your discs.

Luckily, there is a solution to this problem that I've been using since the early 1990s! It requires no computer programming degree or expensive software, and it allows you to create an easy-to-use, fully featured menu system that will run on virtually any PC with Windows. What's my secret? It's HTML—the Hypertext Markup Language—that's used to create the billions of Web pages on the Internet.

In this chapter, I'll provide all the information you need to understand when and why you should add an HTML menu to your discs—as well as why you shouldn't use HTML with every disc you record. I'll discuss the guidelines you should consider while designing your menus. I'll provide you with a comparison of HTML page creation programs, then I'll demonstrate how you can use Netscape Composer to produce your menu system.

Do I Need a Menu?

First, let's consider when a disc needs a menu (and when it's best left as-is): I'll be honest, it takes time to create a good-looking and well-designed menu, and you won't want to take that time and effort with most of the data discs you record. Also, a menu system is unnecessary for some discs: Naturally, you won't need a menu system for a standard audio CD. In this section, I'll consider examples of recording projects that are suitable for an HTML menu system.

Multimedia Material

I always consider an HTML menu whenever I'm sending photos, digital video, or audio files to other people. Just like today's Web pages, your HTML menu can have built-in buttons or links that will display images and video or play sounds and music through the computer's speaker.

tip **Remember, other programs that I've talked about earlier in the book can also create discs that specialize in displaying images and video (such as PhotoRelay, which I cover in Chapter 11 ("Recording Digital Photographs on CD"), and ShowBiz, which I discuss in Chapter 12 ("Making Movies with ArcSoft ShowBiz"). Discs you create with these programs have menus and controls added automatically.**

Internet Content

Talk about a perfect fit! If you need to add Internet content to a project—such as links to your e-mail address, your Web page, or a file that resides on the Internet—any computer with an Internet connection can use these links within your HTML menu.

Text-Based Material

Another type of content that's particularly well suited to HTML display is a long text file; in fact, the Web was originally conceived as a text-based search and retrieval system. For example, a number of commercial discs I have in my software collection use their own HTML-based menus to provide the complete text of historical documents and classic literature, such as Shakespeare's work. Your text can contain formatting, links to other documents, and complete search functionality.

tip **To use all of the formatting and search functions provided by your browser, you'll have to convert text files into documents in HTML format; most Web design programs will do this automatically when you copy text onto the page. You can also convert a document directly into a Web page from within Microsoft Word, which you can use as a submenu in your menu system: Click File and choose Save As, then click the Save As Type drop-down list box at the bottom of the dialog and choose Web Page.**

Off-Site Web Content

I've used this trick a number of times before: If you're recording your Web site to a disc so that you can carry it with you, an HTML menu can provide a front end with direct links to certain pages (which allows others to jump directly to important pages during a presentation without actually having to "click down" from the Web site main menu).

Programs to Download

Figure 14.1 illustrates the download page from my Web site (http://home.mlcbooks.com), where you can try out the shareware and freeware programs I feature in my books. Without an Internet connection, folks couldn't download these files...unless, of course, I create a data disc that contains the same files and an HTML menu system. (In fact, I've done just that.) If you need to provide a way to move programs from a disc to a person's hard drive, an HTML menu is a fast and easy solution.

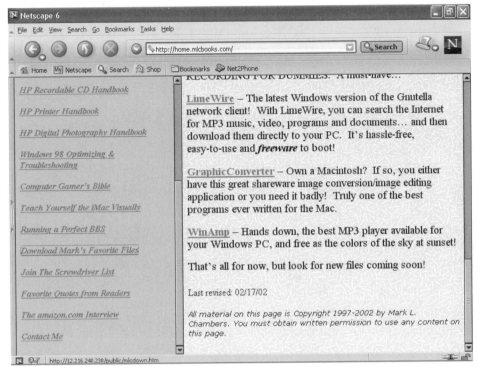

FIGURE 14.1 These download files from my Web site will work well with an HTML menu.

Discs You'll Give to Others

Finally, an HTML menu system is a worthwhile addition to any disc that you plan to distribute to others, because HTML is so compatible among different computer systems (and, if you're trying to create a professional impression, the animation and interactivity you can build into a custom menu will really appeal to the eye). As I mentioned earlier, an HTML menu system is easy for the novice to use, too—just put the index page in the root of the CD, and tell your friends, family, clients, or customers to double-click it. Convenience like that is hard to beat without some sort of programming!

Of course, there are different degrees of necessity here: Just because your project includes one or two digital photographs and a couple of longer text files doesn't automatically mean that you need to go to the trouble of preparing an HTML menu. Just keep the option in mind (and save the files for the menu systems that you do create, because they can be easily reused for other projects).

Adding Menus to a DVD

"Should I add a menu to a DVD?" If your project is a simple data DVD and it follows the guidelines I just outlined, it's probably a good candidate for an HTML menu. In fact, the sheer quantity of files that can be stored on a recorded DVD may make a menu practically a requirement; imagine a computer novice trying to locate a single file among thousands in a folder.

However, don't forget that many of the DVD discs you've learned about in earlier chapters already have a menu system when you record them. For example, if you've recorded a DVD movie disc using MyDVD or ShowBiz, there's no need to add an HTML menu system—these discs can be viewed on your computer already, using programs such as PowerDVD.

tip An HTML menu system is also not a good idea for a rewriteable DVD+RW, DVD-RW, or DVD-RAM disc, because these discs are likely to be reused. There are exceptions to this rule, though—for example, if you rerecord a DVD+RW disc with the same structure and updated files to distribute to software beta testers, you can use the same HTML menu over and over with minor changes.

Why Use HTML?

At this point, you may be wondering why I'm such a fan of HTML and why it has worked so successfully for me (and many of my readers) in the past. (After all, I'm also a programmer and a shareware developer, so I could write a menu system in REALbasic or Visual Basic almost as quickly.) You may be asking yourself, "Can I really write HTML code?

Although HTML is one of the easiest languages to learn on the planet, it's still a computer language, right?"

How about a little reassurance? Before I launch into a full-scale discussion of HTML, I'd like to discuss the details and explain why HTML is so compatible, easy to use, and easy to program.

Easy to Learn

Although it's technically a programming language, HTML is *very* easy to learn, compared with other languages—the Web wouldn't have experienced such phenomenal growth if HTML had been cryptic or hard to understand. In fact, today's sophisticated Web design tools—such as Netscape Composer, which I'll use in the project at the end of this chapter—make it a simple matter to create an attractive menu system with *no* programming work at all! You just run the Web design program, create the menu within that application, and save the result as an HTML page—the rest is done automatically.

Compatible with Most Computers

HTML is a universally recognized standard, so it really doesn't matter whether the person is using a Windows PC, an Apple Macintosh running Mac OS 9 or Mac OS X, or even more technical operating systems such as UNIX, Linux, Solaris, or BeOS. If your operating system has a Web browser, your menu will work.

Large Installed Base

That's a term that software developers often toss around; it means that the vast majority of computers running today already have a Web browser installed, so there's no preparation necessary to use your disc. A person can simply load the disc and double-click to go.

tip **If—through some cosmic chance—someone doesn't have Internet Explorer or Netscape Navigator installed on his or her PC or Macintosh computer, you can even include the distribution version of either browser on the disc for free!**

Did I Mention It Was Free?

Not only are most Web browsers free: You can also find HTML editors and Web design programs included with today's operating systems! For example, a version of FrontPage is likely already installed on your Windows computer. Other HTML editors, such as Arachnophilia (which I'll discuss in a page or two) are free of charge or are very cheap shareware. In other words, you don't have to spend a fortune on development software just to create your own disc menu system.

Familiar Controls

Do you know how to use a Web browser? I'd be willing to bet the answer is yes—in fact, even if you don't know how to use a Web browser, it's simple just to click on a link! This friendly front-end to your disc will help reassure even the most nervous novice.

A Wide Range of Commands

As I've already mentioned, the basic commands of HTML will provide just about all of the functionality that your disc menu is likely to need, without any worry about such cryptic programming terms as *protected memory space* or *dynamic link libraries*. (Two of my favorite examples of the techno-speak that I try to avoid.) The computer's Web browser takes care of all that's necessary behind the scenes, leaving you free to concentrate on the look and feel of your menu.

More to Come

If you are willing to spend money—or you already have a commercial Web design tool— you can add even more functionality to your menu system. Think of all that you've seen on your last visit to the Web, such as Java applets (you can think of them as "miniature programs" that will run directly from your HTML menu), animation, interactive games, and high-quality graphics: With very few exceptions, the rule is that, if it works on a Web page on the Internet, it'll work on an HTML menu, too.

With every passing day, Web designers and software developers are producing more features and functions that you can add to an HTML page. I won't be covering any of these tools in this chapter—after all, this is a book about burning discs—but you can find help and tutorials at your local bookstore or on the Web.

For these reasons, I often say that HTML is "the code for the normal, nonprogramming person"—and that's why it works so well for building disc menus that will work on practically every computer.

As you might have already guessed, the actual process is very simple: After you create the HTML pages that make up your menu system, you add them to your data CD layout. You'll also include any graphics, video, sound, and support files that are required by the menu system; if you're using an HTML editor such as FrontPage, these are automatically saved in the folder you specify.

Designing Your Menus

Before you fire up your HTML editor or Web page design program, pause for a moment and consider what type of menu system you want to create. You should jot down the answers to these questions:

- **Is the appearance of the menu important?** Your menu system can pull out all the tricks of today's Web designers, with animated graphics, video, and sound, or you can choose the Spartan appearance of simple formatted text—which is naturally much easier to produce. As an example, consider a demo disc for a new software product on one hand (which is likely to use plenty of graphics and sound to impress potential customers) and compare that with a clip art disc for distribution to your co-workers (which is strictly utilitarian, as in "I want to download this graphic to my hard drive *right now*, without any eye candy." The demo disc is meant to be explored, whereas the clip art disc simply needs a method of retrieving files.

- **What type of computer will use the disc?** Is your menu likely to be used on the latest hardware, or will it have to run on older computers that may not perform well when trying to display video or high-resolution graphics?

- **Will the contents of the disc be downloaded or viewed inside the browser?** This will determine whether you need any commands for downloading files to the hard drive; for example, a CD with text documents converted to HTML can be viewed directly within the browser.

- **Will you need any Internet connectivity?** If not, your disc menu can be totally self-contained.

tip As a consultant, I have found through personal experience that it's still somewhat dangerous to assume that a connection with the Internet will be available! There's certainly nothing wrong with adding links to your e-mail address or Web page to your menu system; if an Internet connection isn't available, those links simply won't operate. However, if you don't take the extra step of including programs or files on your disc and instead provide a link to those files on your Web page, don't automatically assume that the person will be able to download them...red-faced embarrassment could follow.

- **Are submenus appropriate?** Like a typical Web site, your HTML menu system can provide anything from a simple, one-page interface to a complex hierarchy of submenus; typically, I like to keep things to either a single main menu or a main menu with a single level of submenus. Submenus come in handy when the contents of the disc fall into specific categories: for example, a disc that has company logos and artwork on one submenu page and press releases in Word format on another submenu.

Your next step is to write down each of the functions your menu must provide and where they should appear in your menu. A typical list for a software demo disc might look like this:

- Product features (text shown on the main menu).
- System requirements (text link shown in a submenu).

- Demo program download (download link on the main menu).
- Sales e-mail (e-mail link on the main menu).
- Company Web address (Web link on the main menu).
- Installation instructions and troubleshooting (text link in a sub-menu).

tip Of course, it's impossible to cover Web design in this chapter—there are hundreds of books devoted entirely to the subject! However, there are three basic guidelines that graphic artists and Webmasters use to help steer clear of chaos and keep their menus attractive to the eye. First, if your menu system will use color and graphics, it's a good idea to stay consistent throughout your menu—perhaps your company logo and a single color. Second, remember that "the message is the thing"—your menu system shouldn't draw attention away from the subject, whether it be text, files, or multimedia. Finally, don't forget the "Two Clicks Rule": Make sure that the material you're presenting is no more than two links (or clicks) away from the main menu.

Once you've got the basic design specification on paper—or in Word—it's time to create the menu pages!

HTML Editing vs. Web Design vs. Page Layout

"Hang on, Mark—I've never created a Web page before!" Don't panic, good reader: In this section, I'll discuss what's available and introduce Netscape Composer, which I'll use to create an example menu later in the chapter.

There are actually three types of programs commonly used today to create Web pages: the HTML editor; Web design programs, such as Microsoft FrontPage and Netscape Composer; and page layout/processing programs, such as Microsoft Word.

The Basics: HTML Editing

Because HTML is completely text-based, any text editor can function as a basic *HTML editor*; for example, I often make quick changes to my Web sites using Windows Notepad. No visual "drag-and-drop" page creation here; like any other programming language, HTML is edited using the keyboard. The language uses commands called *tags*, each of which is a line in the file that instructs the Web browser on what to do (for example, which color to use for the background or what size the text should be). Within the editor, you're actually looking at the tags that make up the page; to see the page as it will appear in a Web browser, HTML editors either offer a "preview" mode or automatically launch your Web browser.

To be honest, I would recommend that most PC owners avoid an HTML editor! As you'll see later in the chapter, it's very easy to create a page using a Web design application, which takes a visual approach rather than text-based coding; you don't have to worry about typing a single HTML tag.

On the other hand, if you're already familiar with HTML programming and you'd like to work directly with the commands in an HTML editor, I can highly recommend Arachnophilia (Figure 14.2), a free "CareWare" application that includes professional features such as programming macros, table support, and a spell checker. You can download this great HTML editor at www.arachnoid.com/arachnophilia/index.html.

The Visual Approach: Web Design

Next, let's consider Web design applications such as Microsoft FrontPage and Netscape Composer: These programs offer a much easier-to-use "drag-and-drop" approach to page creation, so you won't have to edit any actual HTML code. Once you've added text, inserted graphics, and set up your links, the application automatically generates the HTML file for you. Definitely the way to go for most of us!

Because these applications are meant to help those with no design experience, a Web design application will likely have wizards to walk

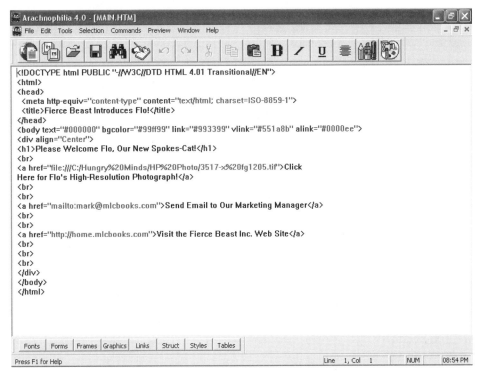

FIGURE 14.2 An HTML editor such as Arachnophilia displays "raw" HTML code in text format.

you through the page creation process, as well as a number of ready-made templates and designs that you can use or modify for your menu system. You won't need to know a single HTML command to use one of these programs, and once you're familiar with the way your Web design program works, you can produce a quality menu page for a disc project in a few minutes.

By the way, if you don't already have a Web design program (and you'd rather not spend any money on a shareware or commercial application right now), visit www.netscape.com and download the latest version of the Netscape Web browser—you'll also install a free copy of Netscape Composer, the company's easy-to-use Web design program.

The Alternative: Page Layout and Word Processing

Today's word processing and graphic page-layout programs have branched out into Web design, as well, including such workhorses as Microsoft Word and Adobe Pagemaker. Although these programs don't have the full range of specialized features and functions of a dedicated Web design application, they do have one big advantage: You're likely already familiar with them! For example, Figure 14.3 illustrates one of the ready-made Web page templates that come with Microsoft Word.

The HTML menu pages produced by these applications will vary quite a bit in functionality and appearance, but generally the result will be more than adequate for all but the most demanding multimedia menu system.

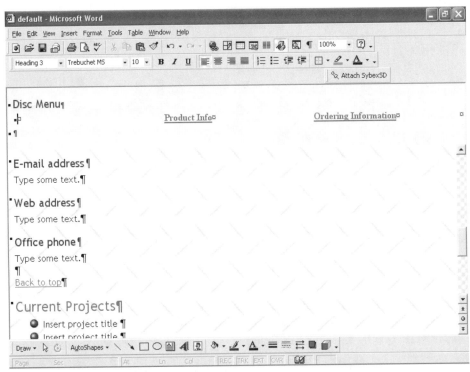

FIGURE 14.3   This simple HTML page was created in seconds, using Microsoft Word.

tip **If you're unsure whether your favorite page layout or word processing program can produce Web pages, check to see whether it can save (or export) documents in HTML format.**

Project: **Building an HTML Menu**

Let's create an example single-page menu for a fictitious cat food company, Fierce Beast Incorporated. The disc will contain a digital, high-resolution photograph of the company's new spokes-cat that can be displayed, as well as links to the company's Web site and an e-mail link to the company's marketing contact. I'll be using Netscape Composer under Windows XP.

Requirements

- An installed copy of Netscape Composer
- A digital photograph

✔ Follow these steps:

1. Click Start and choose All Programs | Netscape 6.2 | Netscape 6.2, which runs Netscape Navigator.

2. Within Navigator, click File, select New from the menu, and click Blank Page to Edit. The Composer screen you see in Figure 14.4 appears.

3. Let's begin by changing the rather plain white background and adding a little color. Click Format and choose Page Colors and Background, and Composer opens the dialog you see in Figure 14.5.

4. Click Use Custom Colors and click the Background button, then select a lighter color (which will keep our text easy to read) from the samples. Click OK to accept the color change.

FIGURE 14.4 Netscape Composer, ready for duty.

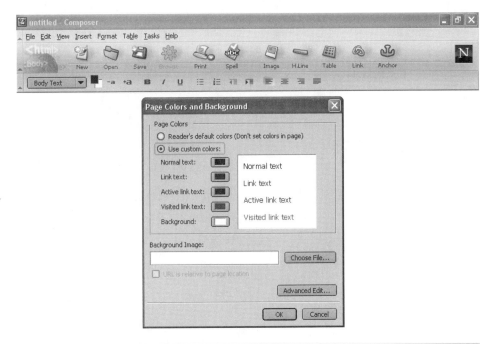

FIGURE 14.5 Selecting a suitable color for our background.

5. Next, we need a title: I'll type *Please Welcome Flo, Our New Spokes-Cat!* Like a word processor, you can simply type text directly into Composer.

6. The program also allows you to edit using the familiar click and drag, so select the text, click the Paragraph Format drop-down list box, and choose Heading 1. Click Format, choose Align, and click Center on the pop-up menu to finish our title (Figure 14.6).

7. Our first link will be to the photograph FLOPIC.JPG, which will be included in the root directory of the CD. Press Enter twice and click the Link button on the toolbar; Composer displays the dialog box shown in Figure 14.7. Type the text that will appear on the menu in the first box.

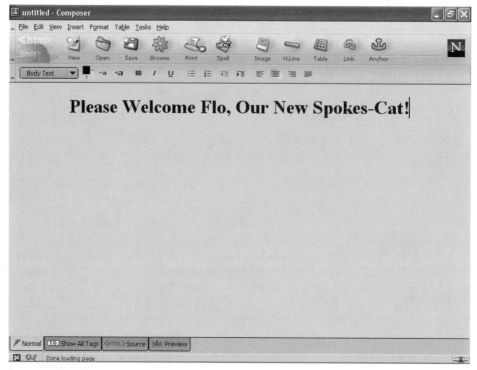

FIGURE 14.6   Our title is now in place.

FIGURE 14.7 The Link Properties dialog.

8. Now we must "connect" the link to the file on the disc. Click the Choose File button to locate the file. Highlight the filename and click Open. Click OK to create your first link (Figure 14.8)!

   tip **If your image doesn't show up in the Choose File dialog, Composer is looking for only HTML files; make sure that you choose the All Files view in the Files of Type drop-down list box.**

9. Because the image will actually be stored on CD (and not on the computer's hard drive), right-click the photo link and choose Link Properties. Click in the Link Location field, erase the current contents (which point to your hard drive), and type the name of the file—in our case, FLOPIC.JPG. Click OK, and the computer will now look for the file in the same location as the HTML file. (Remember, both of these files will be included in the root direc-

FIGURE 14.8 You did it! That's a link to your photo.

tory of your data CD layout, so don't panic because the link doesn't work at this time.)

10. Next, add the e-mail link. Press Enter twice, click Insert, and choose HTML to display the dialog shown in Figure 14.9. Type this text into the box:

    ```
    <A HREF ="mailto:mark@mlcbooks.com">Send Email to
    Our Marketing Manager</A>
    ```

11. As you might have guessed, the line you just typed was an actual HTML tag—I wanted to give you a taste of what the language looks like. Change the e-mail address (the part immediately following the mailto: that appears within the quotes) to your e-mail address and click Insert, and the link appears (Figure 14.10).

FIGURE 14.9 Typing an HTML tag into Composer.

FIGURE 14.10 You've added an e-mail link to your menu.

12. Finally, we need to add a link to the company's Web site. Press Enter twice and click the Link button on the toolbar to display the Link Properties dialog again; type the text for the Web page link in the Link Text box.

13. To specify the Web address, click in the Link Location field, and enter the full address—for example, a link to my MLC Books Online site would be `http://home.mlcbooks.com` (note that the `http://` part is required). Click OK, and Figure 14.11 illustrates our completed menu!

14. Click File and choose Save As to save the page. Composer displays the dialog you see in Figure 14.12: Type the text that will appear in the browser's title line, and click OK.

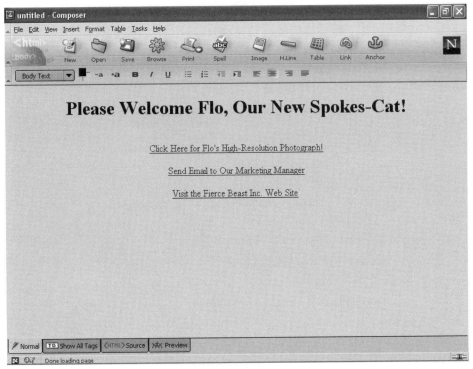

FIGURE 14.11 Our basic disc menu is ready to save.

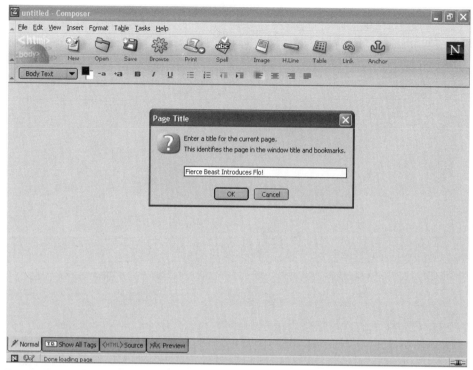

FIGURE 14.12 Entering a title for the completed page.

15. From the Save Page As dialog box, select the folder where the menu should be saved. Click in the File Name text box, enter a short name, such as MAIN.HTM, and click Save.

You're ready to burn a test disc! Open HP RecordNow or Easy CD Creator and create a disc layout that includes the menu files you created—in this case, MAIN.HTM—and any files that will be downloaded or displayed—in this case, FLOPIC.JPG—in the root directory of the disc. (This is very important, because recording these files in other locations—like folders *within* the root directory—will cause your links to fail.)

Once you've recorded your disc, you can test your menu. Load the disc and run your Web browser, then click File and choose Open (in Internet Explorer) or Open File (in Navigator). Locate and double-click on the MAIN.HTM file on the CD, and click on all the links to make

FIGURE 14.13 That's some cat!

sure they work. Figure 14.13 illustrates our completed menu (and Flo herself!), as displayed in Navigator.

Naturally, this project only scratched the surface of what's possible—we didn't add any graphics or sound, for example—but it also showed you just how easy it is to create a basic menu. As you become more familiar with your Web design software, you'll be able to enhance the look of your menus and expand the functionality of your disc projects.

Summary

This chapter introduced you to HTML and demonstrated how an HTML menu can add an attractive appearance and the convenience of a point-and-click interface to your CD projects. I discussed which types of discs will benefit from a menu system, as well as the basic concepts of menu design and the HTML applications that you can use to create your pages.

chapter

15

Converting Albums and Cassettes to Audio CDs

In This Chapter

✓ Connecting your stereo and computer

✓ Project: Transferring a Cassette to MP3 Files

Are your rare and precious vinyl albums "dying a slow death" from scratches and warping? Are your home-recorded cassettes losing their range and quality? Or (and this shows my age), do you have 8-track tapes that are close to extermination?

If you're like me—an audiophile with several dozen albums and cassettes of irreplaceable music and vocals—you'd prefer to have that material on audio CDs, instead. But how can analog data be transferred to digital format and recorded for posterity?

his final chapter shows you how to use the Spin Doctor portion of Roxio's SoundStream—one of the programs included with Easy CD Creator 5—to make the jump from analog to digital. You'll learn about the equipment you'll need, and I'll demonstrate how to transfer the audio from an album to an audio CD.

A Word of Caution

Before you read any further, take note: *Transferring analog to digital should be done only for those recordings you can't already get on audio CD!* In other words, dear reader, it's just plain ridiculous to spend your valuable time and effort copying your old Beatles albums to an audio CD. Why? Those albums are almost certainly available already on audio CD, and no matter how good a job you do in transferring the music, it will always sound better when professionally mastered from the original recordings. (Plus, you're likely to get liner notes.)

With this in mind, consider an analog-to-digital transfer a last resort for the truly *irreplaceable* audio in your collection: tapes you've made of your family, for example, or your audio diary. Or that bootleg recording of Ella Fitzgerald and Frank Zappa jamming together live. (Come to think of it, send me a copy of that one, too.)

What Do I Need?

To transfer analog-to-digital, you'll need the following equipment (besides the SoundStream software):

- **A stereo system that can play the original media.** Your stereo's amplifier, preamp, or tuner needs standard Audio Out jacks that will allow it to connect to your computer.
- **Your computer's sound card.** Virtually all computer audio cards now have Audio In jacks, and any card should be able to record CD-quality music.

- **The proper cables.** Ah, here's the rub. Unfortunately, many computer sound cards on the market today have proprietary connectors—often called *dongles* (and usually with irritation)—that are required to connect standard audio cables.

Due to the wide range of stereo hardware and the equally wide range of computer sound card connections, I can't provide any specific instructions for connecting the cables; luckily, your sound card manual should include this information (as well as an explanation of which audio jacks you need to use).

Project: **Transferring a Cassette to Audio CD**

Without further ado, let's copy a cassette to separate MP3 files, then to an audio CD—in this case, a live performance of a local band that I captured from the DJ's sound board. (It's not likely that I'll encounter this band in a record store, because they split up over a decade ago without ever releasing anything except a handful of demo singles!)

Requirements

- Original audio media—for this project, a cassette tape
- An installed copy of Easy CD Creator 5, which includes Sound-Stream
- An installed copy of HP RecordNow (or other recording software)
- A stereo with a cassette deck and cables to connect to the computer's sound card
- A blank CD-R disc

✔ **Once you've connected the cables between your computer and your stereo, follow these steps:**

1. Run SoundStream by clicking Start and choosing All Programs | Easy CD Creator 5 | Applications | SoundStream. The program's main menu appears, as shown in Figure 15.1.

2. Click on the Option Drawer button—the button at the bottom of the SoundStream window with the arrow) to open the drawer, and click Spin Doctor to display the dialog shown in Figure 15.2.

3. Choose Sound Cleaning, and click the Tape preset button to allow Spin Doctor to choose the best settings for Sound Cleaning and Pop/Click Removal.

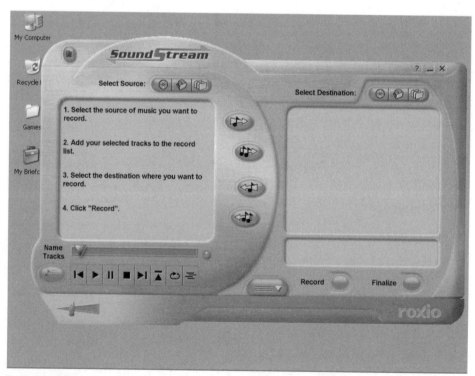

FIGURE 15.1   Roxio's SoundStream can transfer any analog audio signal to an audio CD.

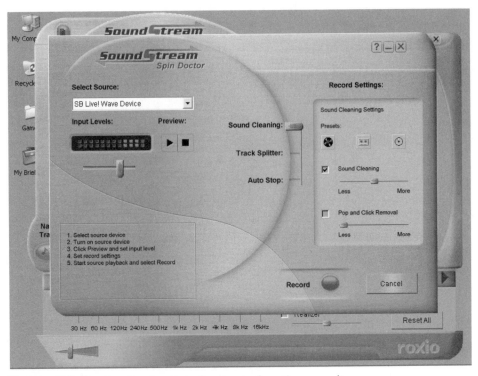

FIGURE 15.2 Configuring Spin Doctor before the recording.

tip If you have a single sound card in your PC, leave the Select Source drop-down list box set to the default—however, if you have more than one sound card or sound input device, you may have to use this option to choose the proper input/record source.

4. Choose Auto Stop, and choose either Manual Stop—where you have to stop the recording by hand—or Auto Stop after 3 to 4 seconds of silence. Personally, I use Manual Stop (I like to be in complete control), but as long as there are no songs with long periods of silence, Auto Stop should work fine.

5. OK, it's time to check the audio levels. Turn on the stereo and begin playing the cassette—it's typically a good idea to set the stereo amplifier volume about one-quarter to halfway.

6. Click the Preview button, and click and drag to adjust the Input level slider until the loudest passage of music lights the entire green portion of the level meter. Note that you do not want the peak portion of the music to move the level meter into the red zone for more than a split second! When the level is properly set, click Stop.

7. Now you're ready to start the actual recording process. Click Record to display the dialog shown in Figure 15.3. I recommend recording the music in File format, which allows you to check the quality of the recording later, without potentially wasting a blank CD-R disc. (Plus, this also allows you to add the title, artist, and track names before you burn the disc.) Click File.

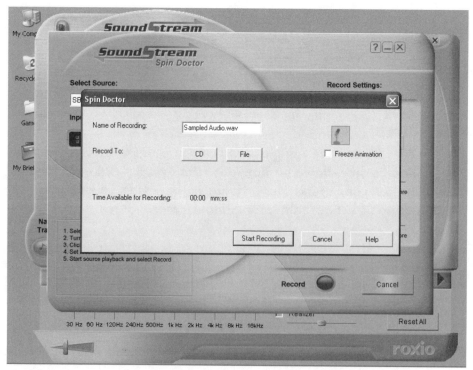

FIGURE 15.3 It's a good idea to record your tracks to MP3 first.

8. Click the Save As Type drop-down list box on the Save As dialog (Figure 15.4) to specify WAV or MP3 format, and select the folder where the tracks should be stored (or click New Folder to add a new folder on your hard drive). When you've navigated to the proper spot, click Select Folder.

9. Rewind the cassette and start it playing, then click on Start Recording. If you chose Auto Stop, Spin Doctor will create separate tracks in the folder you specified.

10. Once the last track on the cassette has finished, click Stop and click OK to return to Spin Doctor.

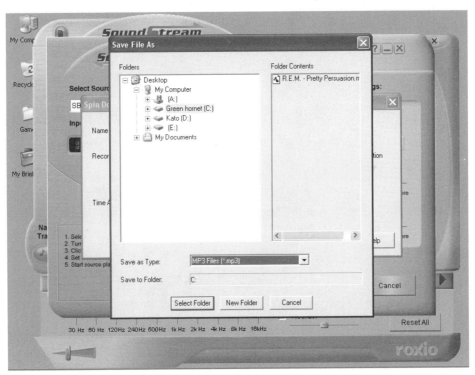

FIGURE 15.4 Specifying a location and format for your sound files.

Congratulations! That aging cassette has been transferred to clean, pure digital satisfaction. However, before you toss it into the closest trash can, you'll want to review each track using an MP3 player, such as Windows Media Player or WinAmp, to make sure that the recording went well. Once you're satisfied with all of the tracks, use HP RecordNow or Easy CD Creator to create a standard audio CD with the MP3 files you've recorded.

Summary

In this final chapter, I demonstrated how to use SoundStream to transfer audio from older media, such as albums and cassettes, to digital MP3 format, where they can be enjoyed and—optionally—recorded to an audio CD.

A

Hewlett-Packard Technical Support's Frequently Asked Questions

The questions and answers in this appendix were collected by the Hewlett-Packard technical support staff at Customer Care Centers around the world.

tip **Additional information is always available online at the Hewlett-Packard CD Writer home page (http://www.hpcdwriter.com/).**

How much can I fit on a 74-minute CD?

For audio, the answer is simple enough:

HP RecordNow

Using this software, you should be able to place 650 MB of data on the disc. You can store 74 minutes of audio on a 74-minute CD. To maximize the storage space on an audio disc, use DAO recording mode (Disc at Once), which removes the 2-second gaps between tracks.

HP DLA

The capacity of an HP DLA disc depends on whether you are using a CD-R or CD-RW disc. CD-R, when formatted with DirectCD, usually will have between 550 and 600 MB left after formatting. A CD-RW disc will generally have between 500 and 550 MB left after formatting.

Why do my recorded audio CDs hiss and pop?

This is caused primarily by the source drive used to extract the audio tracks for burning. Many older CD-ROM drives were not designed to perform the audio extraction technique required by today's CD and DVD recorders at a fast enough speed. Errors normally not audible in the source content often surface during this process. To test and verify the issue, use your recorder as both the source and the destination drive for an audio CD burn. On the resulting disc, the audio quality should be corrected. If this resolves the issue, you may continue to use the recorder as your source drive, or you may wish to investigate a new CD-ROM for your computer, capable of faster audio extraction.

Why can't I read my recorded CDs on another computer?

There are a few different issues that can affect what we call *disc interchange,* or reading a burned CD on another computer system or CD device.

CD-R vs. CD-RW Media

CD-R discs are recommended for any CDs you wish to take to another computer system. CD-RW can be used, but some older CD-ROM readers will have problems accessing the information on these discs. If the CD-ROM reader has a MultiRead icon on the front plate, that drive will be able to read RW media.

Using Roxio's Easy CD Creator or CD Copier Deluxe

These two burning applications write to the CD in the standard track format, adding a table of contents to the end of the session. This method of writing is most widely recognized by other CD devices and will increase the chance for successful disc interchange.

Even with this method, however, the preceding information about CD-RW media still applies. Using Easy CD Creator or CD Copier Deluxe with RW media may still create difficulties reading the information, unless the reader device is MultiRead-capable. Using CD-R with this software, however, affords you the best chance of reading your created CDs on another computer and is the recommended method.

How do I clean my HP CD or DVD recorder?

HP CD-Writer and DVD-Writer drives require no maintenance or cleaning. If you simply keep your discs clean, you will prevent most problems.

Do not attempt to use the CD-ROM cleaning CDs that use a small brush to sweep dirt off the laser mechanism.

How do I create discs for other operating systems?

The process for setting up the software is simple. Using Easy CD Creator, click File on the menu, select CD Project Properties, then select ISO 9660 from the drop-down menu. This will set the software to burn the CD in the most universal format available. Even with the disc burned properly, there are some things to keep in mind:

- The naming structure of any files you will be burning must meet the 8.3-character naming convention set forth in DOS. Additionally, if you select an option other than MS-DOS 8.3-character naming, the software will allow you to use more types of characters (*&%$#), but this will make the CD less likely to be readable on another operating system.
- Remember that, even though you can create a disc that another system can read, the code within the files may not make sense to a given operating system. For example, if you create a CD for a Macintosh computer containing a program written for Win-

dows, the Mac's CD-ROM drive will be able to see it, but the Mac will not be able to run the program.

Why are files that are copied from a DVD or CD to the hard drive assigned the Read Only attribute on the hard drive?

This will always be the case with optical media such as CD-R discs, because they are read-only media. This is actually one of the more important features of optical technology, in that data is virtually incorruptible once it's written to disc. It cannot be accidentally overwritten or erased when the disc has been closed. The most straightforward approach to resolving this problem is to change the attributes of the files after copying them to the hard drive. This procedure is accomplished by selecting the files individually or as a group, right-clicking the mouse button, and choosing Properties. The dialog box that appears enables you to remove the Read Only attribute.

I was formatting a disc with DirectCD and the process was interrupted . . . now I can't use the disc! Can I fix this?

If power is lost while a CD-RW disc is being formatted, the disc will become unusable in any application, and it will fail if another format is attempted.

To solve this problem, use the Easy CD Creator Erase feature to erase the disc fully, then try the format operation again.

Can I play an audio CD on either my internal HP recorder or my CD-ROM if my sound card has only one input?

Yes. Use an audio "Y" cable connector. Some of these connectors require that a small switch be installed in the system case. Toggling the switch will enable you to select either drive for playback. Some newer sound cards allow two devices to be connected, so check with your sound card manufacturer before buying a Y cable connector.

Some sound cards have more than one audio connector, but they're not suitable for connecting two devices—the extra connector is only to accommodate different styles of audio cables. Attaching more than one CD-ROM/CD-Writer to these types of sound cards is not recommended. Check with your sound card manufacturer before connecting more than one drive. For example, some versions of the Sound Blaster Live sound card have an SP/DIF digital input, which is not compatible with the output from a recorder.

B

Tips on Buying a Recorder

If you're like most readers of this book, you probably already have your DVD recorder: Either it came with your computer as standard equipment, or you've already bought it, and now you're ready to install it. (You'll find installation guidelines and procedures in Chapter 2, "Installing Your DVD Recorder.")

However, if you don't fit into one of these two categories and you haven't bought your recorder yet, you'll find this appendix a valuable list of features and tips that you can use while shopping for the right drive.

Picking the Features You Need

This section will help you determine which features you should look for when shopping for your drive.

Internal or External?

For most of us, an internal DVD recorder is likely the right choice. A number of reasons have made internal drives popular:

- An internal drive doesn't require a separate power supply.
- It doesn't take any additional space on your desktop.
- An internal recorder is typically cheaper than an external model.

On the other hand, if you need these specific advantages, an external recorder may be right for you:

- External drives are much easier to install, because you don't have to open your computer's case.
- Some computers—such as a typical laptop or an "all-in-one" computer, such as the Apple iMac—may not have internal bays for expansion (or you may have internal drive bays, but they may be all used for other devices).
- An external drive can be carried with you on trips.
- External drives are easy to share between computers—for example, a small office with computers that aren't already hooked to a network.

To save money—and if you don't need to transport your drive or if your computer can't be expanded—I usually recommend an internal recorder.

Comparing Drive Speeds

In general, today's DVD recorders will record all of the common formats that I've discussed in the book: audio, data, mixed-mode, CD-Extra, Video CD, and UDF/packet-writing. However, how fast they can record a disc is a different matter! The more discs you record, the more that extra speed will save you time.

Drive reading and recording speeds are measured by something I call the *X factor*—for example, a drive advertised as 12x/10x/40x means the drive has a 12x recording speed, a 10x recording speed with rewriteable discs (which is a significantly slower process), and a 40x read-only speed. The faster the X factor, the better the drive's performance (and typically, the more expensive). CD-ROM and DVD-ROM read-only drives usually carry only one X number, which refers to the speed that a disc can be read.

So what does the X factor stand for, anyway? It's a multiplier of the original transfer rate for the first single-speed CD-ROM drives that appeared in 1980, which was 150 KB per second. To illustrate, a 40x CD-ROM can transfer data at a whopping 6,000 KB per second.

The Advantage of Burn-Proof Recording

I've mentioned burn-proof recording earlier in the book, but it's worth a reminder here: If at all possible, *make sure that your new drive has this feature*, because burn-proof recording will eliminate buffer underrun errors and prevent you from wasting blank discs. Additionally, a drive with burn-proof recording can be used while you're working on other tasks—for example, I've often recorded discs while working on book manuscripts in Microsoft Word, and because of this feature, I've never encountered a recording error (even when Word was making heavy use of the processor and hard drive when loading, saving, and virus checking documents).

Luckily, burn-proof recording is now a common feature on the latest drives, and you no longer have to pay a king's ransom and buy the most expensive drive to enjoy the benefits of burn-proof recording.

Choosing a Larger Data Buffer

Your recorder's data buffer (sometimes called a *cache* or *internal RAM*) is designed as a "storage area" for data that has already been read from your hard drive but isn't ready to be recorded to the disc yet. A buffer is necessary because today's hard drives can read and transfer data much faster than your recorder—therefore, the buffer helps ensure that the entire recording process is as efficient as possible by providing your recorder with a steady supply of ones and zeroes to write to the disc!

However, a closer inspection of the specifications for today's drives shows that many recorders offer far more data buffer space than others—the storage space can range anywhere from 512 KB of buffer memory on the low end to 1, 2, or even 4 MB on better drives. Just remember that the larger the buffer your drive has, the better will be your drive's performance—particularly at its maximum recording speeds or if you're using your computer for other tasks while you're recording.

Shopping for Audio Controls

If you're an audiophile and you've added a great-sounding expensive speaker system to your PC, I heartily recommend that you consider investing in a recorder that has a full set of audio CD drive controls (along with the ubiquitous volume control and headphone jack that appear on every drive). You'll find it much easier to use these buttons, rather than clicking on the buttons on your audio CD player program!

Audio CD drive controls usually include track-forward and track-reverse buttons, a pause button, and a play button, mirroring the controls on a standard audio CD player.

tip **If you're wondering where to find all of these features and specifications on the drives you're considering, visit each manufacturer's Web site—most drive manufacturers now provide a complete description and specification sheet on each model they make. Use the Google search engine at www.google.com to search the Web for specific drive models; you'll be amazed at the range of information you find, including various newsgroup messages and comments from other drive owners. You can also find details—and often, even compare drive features on one screen—on www.buy.com, www.pricewatch.com, or www.computershopper.com. Finally, don't forget that computer magazines will feature hardware comparisons, so check your favorite magazine's Web site for any reviews that cover the models you're considering.**

The Importance of Technical Support

It's a common mistake, and most computer owners shopping for new hardware have made it: neglecting to evaluate the technical support available for your new recorder (or any hardware device) before you buy it. (Believe me, this is a tough lesson to learn *after* the fact.) Certainly, most manufacturers appear to offer the same range of technical support services—usually including a Web site, telephone support, and updated drivers—but these services are definitely not "created equal."

For example, consider the questions you should be asking about the telephone tech support that's provided with the drive you're evaluating:

- Is telephone tech support available at all, or is all tech support handled by e-mail?
- Is a toll-free phone number available for voice support?
- Is voice support free, or will you be charged for it?
- Can you get support 24 hours/7 days a week, or will you have to wait for business hours during a weekday?

When you visit the company's Web site, check for the availability of updated drivers and firmware updates. Are they released regularly, or is the last driver listed with a date of 1999? Other Web criteria to check include:

- Does the manufacturer provide a "bulletin board" message area for owners to converse together online?
- Are there FAQ (short for *Frequently Asked Questions*) files you can read with common problems and solutions?
- Is there a knowledge base you can search for answers to your technical questions?
- Are specifications and user manuals provided for downloading?

Remember, it's just as important to "peek behind the curtain" when it comes to technical support as it is when comparing hardware features.

Tips for Buying Your Recorder Online

Most of us are familiar with the advantages of buying locally—immediate gratification and easier returns for defective hardware—but you're practically guaranteed to get a better price by buying your new drive online. I'll end this appendix with the top four tips I provide my readers when they ask about buying a drive online.

- **Be cautious about buying refurbished hardware.** I know that refurbished—read that as "used and repaired"—hardware is very inexpensive, compared with the same hardware bought new, but make sure you check the return policy for a refurbished recorder, and buy refurbished components only from a reputable company. You have no idea what broke the first time, and you'll be taking a chance if you can't return a "lemon" that looked like a good deal.

- **Compare prices before you buy.** This sounds like a given, but many folks buy from the very first Web site they visit (or from a site with a name they recognize). Don't get me wrong—building an ongoing relationship with a Web store you trust is a good thing, and I have a number of favorite stores that I always visit—but would you buy something *locally* without checking other stores for their prices? The same holds true online; for example, I recently saved myself over $100 by doing 20 minutes of comparison pricing on a new computer. As a starting point, I heartily recommend www.pricewatch.com, where you can search for a specific drive and see at a glance the range of prices available from popular Web stores.

- **Don't buy online without a secure connection.** The chance of a hacker obtaining your credit card number and personal information is relatively slight, but it is certainly possible. Therefore, I buy only from stores that offer a secure, encrypted connection. When information is encrypted, a hacker has a much harder time intercepting your data as it's being transmitted to the Web site. Both Internet Explorer and Netscape Navigator tell you whether a site has an encrypted (or secure) connection: Check to make sure that your browser is displaying a closed padlock icon on the status bar at the bottom of the screen. If the connection isn't secure and you're being prompted for your credit card number or personal information, find another store!

- **Avoid high shipping costs.** Many online stores now default to hideously expensive next-day or overnight shipping for your drive—usually, you can save up to half of those shipping expenses if you wait a day or two more for your recorder and pick ground shipping! Pay close attention to what you're paying *before* you click the Buy button.

appendix

C

Using Musicmatch Jukebox

*Although most PC owners are familiar with MP3 files—
what they are and how to get them—there are a number of
folks I've met who have only listened to audio CDs on their
computers. If you've never played or extracted (or "ripped")
tracks from an existing audio CD in your collection, then
this appendix will guide you through both procedures in two
projects; we'll use Musicmatch Jukebox, my favorite MP3
player/utility/toolbox program!*

 tip You can get a free copy of the Basic version of Jukebox
by visiting the Web site at www.musicmatch.com.

Project: Extracting MP3 Files from Existing Audio CDs using Jukebox

Musicmatch Jukebox makes it easy to rip a track from an audio CD
using your DVD recorder; when the process has finished, you'll
have the track in MP3 format on your hard drive.

caution   You can only extract tracks from audio CDs that you own—for example, to create a compilation CD of your favorite songs from a single artist. If you didn't buy an audio CD, you're violating copyright law by ripping tracks from it.

Requirements

* An installed copy of Jukebox
* An audio CD

✔ **Follow these steps to rip MP3 files:**

1. Run Jukebox (under Windows XP, click Start->All Programs-> Musicmatch->Musicmatch Jukebox) to display the program's main window, as shown in Figure C.1.

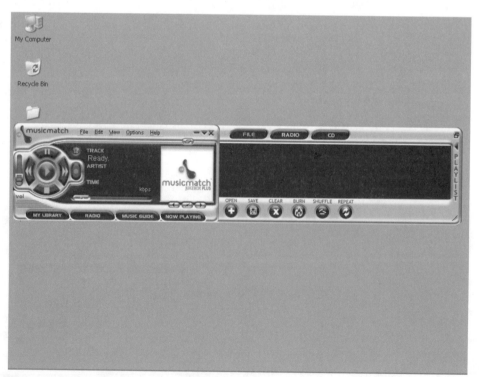

FIGURE C.1   The Jukebox window.

2. Load the audio CD into your DVD recorder, and Jukebox automatically displays the track names and begins playing the CD (Figure C.2). Since we don't want to listen to the disc right now, click the Stop button on the circular control panel to stop the playback (it has a square icon, just like an audio CD player).

3. Next, it's time to pick the proper settings for our MP3 recording—in this case, I want to rip Track 1, "Movin' In" (in my opinion, one of the best songs Chicago ever performed). Click the Options menu and choose Recorder->Settings to display the dialog you see in Figure C.3. Choose MP3 recording format, CD Quality (128 kbps) and make sure that the Recording Source is set to your DVD recorder. Click OK to save any changes that you've made.

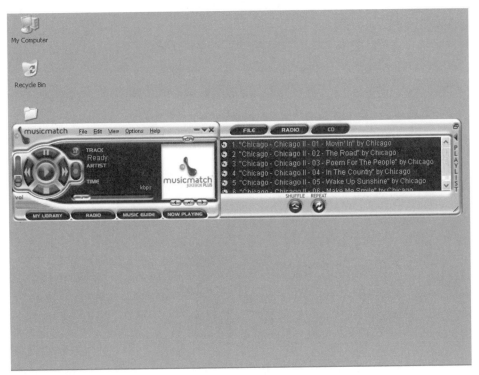

FIGURE C.2   Jukebox begins playing my Chicago II CD.

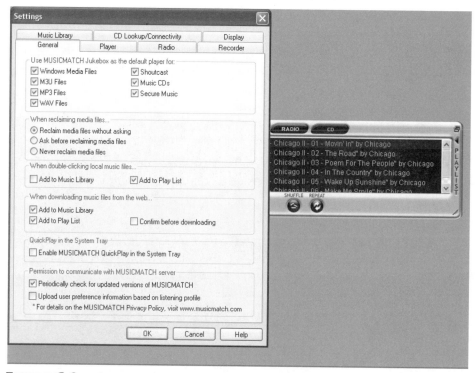

FIGURE C.3   Setting recording options for Jukebox.

4.   Now it's time to start the actual extraction! Click the red Recording button on the circular control panel to display the Recorder panel (Figure C.4). To mark a track for recording, leave it checked—clear the checkbox next to each track you want to skip. In this case, we want to make sure that only the first track is checked.

tip   **By default, Jukebox marks all the tracks on the disc for recording—if you only want one or two tracks, however, you can click the None button to uncheck all of the tracks, and then click the checkboxes for just the tracks you want.**

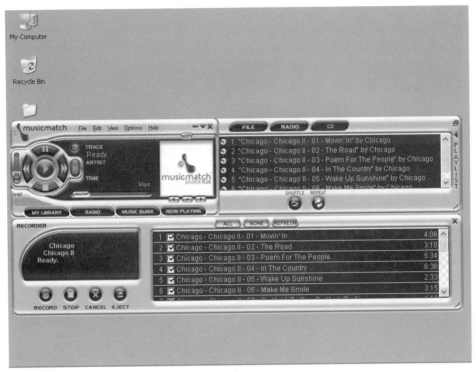

FIGURE C.4   Jukebox is now in "Record" mode.

5. Click the red Record button on the Recorder panel. (If this is the first time you've ripped tracks with Jukebox, the program performs a short test and calibration procedure, then continues.) You can watch the progress of the extraction—Jukebox displays a progress bar next to the track name (Figure C.5).

tip   **If you're using Windows Me or Windows XP, Jukebox stores the MP3 files you've downloaded in the My Music folder in the My Documents folder. A separate folder is automatically created with the artist and album name to keep things organized. Neat!**

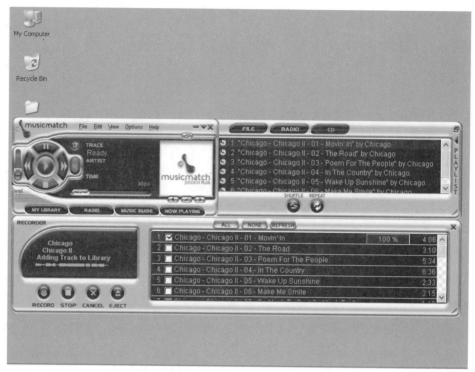

FIGURE C.5   The extraction progress bar lets you monitor the ripping process.

Project: **Extracting MP3 Files from Existing Audio CDs using Jukebox**

Now that I've demonstrated how to rip MP3 files, let's use Jukebox to listen to them.

tip If you're using a PC running Windows Me or XP and you don't have Jukebox installed, you can also use Windows Media Player to listen to MP3s; just double-click on the file to launch Windows Media Player.

Requirements

- An installed copy of Jukebox
- MP3 files on your hard drive

✔ **Follow these steps to listen to MP3 files:**

1. Run Jukebox and open an Explorer window.

2. Drag the MP3 files you want to hear from the Explorer window and drop them in the Jukebox Playlist pane at the right, and they'll be automatically added. Although Jukebox will automatically begin playing the first song in the list, you can continue to add MP3s and they'll appear at the end of the Playlist.

3. Click and drag the Volume slider control to set the desired volume. You can also use the Previous, Next and Pause keys to move through the Playlist or pause the playback while you're away from your PC.

tip **Alternately, you can simply double-click on an MP3 file in Explorer, which automatically launches Jukebox and plays the song—however, use the procedure I've demonstrated to add multiple tracks and create a Playlist.**

Glossary

A

Adapter card A plug-in computer circuit board that you install to add functionality, such as a FireWire port adapter card.

Audio cable All DVD and CD recorders have a connector for an audio cable that runs from the drive to your computer's sound card—this cable is used when you're playing an audio CD using your recorder.

B

Binary Computers use the binary language to store data (including the data on DVDs and CDs) and to communicate with each other. Binary data has only two values, zero and one.

Bootable CD A CD-ROM containing all of the operating system and support files necessary to boot a computer without a hard drive. Bootable CD-ROMs use the El Torito format.

Burn-proof recording A feature offered on most new recorders that allows the drive to pause during recording when the data transfer rate falls below an acceptable level—for example, when your computer is busy opening a large spreadsheet. A drive with burn-proof recording will not return "buffer underrun" errors.

Byte A single character of text stored by your computer (on a DVD, in RAM, or on your hard drive).

C

Case Your computer's metal cover—when you install additional hardware, you can remove the case by unscrewing it or opening a hinged door.

CD-Extra A specialized mixed-mode disc that contains both audio tracks and a data track; unlike a standard mixed-mode disc, a CD-Extra disc can also be played in a regular audio CD player (the audio track is recorded first on the disc).

CD-R Short for *Compact Disc-Recordable*. The first CD-recording technology that was generally available to the public; a CD recorder can store computer data and digital audio on a CD-R by creating a pattern on a layer of dye with a laser beam. Unlike a reusable CD-RW disc, a CD-R disc can be recorded only once.

CD-ROM Short for *Compact Disc-Read-Only Memory*. A CD-ROM is a CD with computer data recorded for use on a computer.

CD-RW Short for *Compact Disc-Rewriteable*. A recordable CD technology that uses an amorphous crystalline layer—the crystalline layer can be changed over and over, so the disc can be erased and rewritten.

cDVD A specialized Video CD format supported by the MyDVD recording program. cDVDs can hold digital video and can be displayed on most computers running Windows.

Chapter DVD movie discs are usually separated into sections called *chapters* that allow the viewer to jump from one major section of the film to another.

Crystalline layer The layer of amorphous crystals in a CD-RW disc that darkens or discolors when struck by a recorder's laser beam. Unlike the CD-R disc's dye layer, the crystalline layer in a CD-RW disc can be "reset," and the disc can be used again.

D

Data buffer Internal memory (similar to your computer's RAM) used by your CD or DVD recorder to store data from your hard drive until it's ready to be recorded. Generally, the larger the data buffer, the fewer errors your drive will encounter while recording discs.

Defragmenting The process of optimizing your hard drive by reading the files it contains and rewriting them in contiguous form, one after another. Files written contiguously can be read faster and more efficiently—a good idea when burning files from your hard drive to a CD or DVD.

Digital audio extraction The process of copying tracks from an existing audio CD to your hard drive as WAV or MP3 digital audio files. Extraction is often referred to as *ripping*.

Disc-at-Once When recording in Disc-at-Once mode, your drive writes the entire disc at once without turning the recording laser off between tracks. Older CD recorders may not be able to record discs in Disc-at-Once mode.

Disc image A file stored on your hard drive that includes all of the information necessary to create a CD or DVD. You can burn copies of a disc directly from a disc image.

Digital video Full-motion video stored on your hard drive (usually in AVI, MPEG, or MOV format). Digital video can be recorded to CDs and DVDs and viewed on your computer or your DVD player.

DVD Short for *digital video disc*. A DVD disc can hold anywhere from 4.7 GB to 17 GB. DVD discs can hold computer data, digital audio, and movies in MPEG format.

DVD+R/DVD-R Standard DVD recording formats that typically store 4.7 GB on a single disc. DVD+R and DVD-R discs cannot be rewritten.

DVD-RAM A standard DVD recording format that can store up to 9.4 GB on a single disc. DVD-RAM discs are rewriteable.

DVD+RW/DVD-RW Standard DVD recording formats that typically store 4.7 GB on a single disc. DVD+RW and DVD-RW discs are rewriteable.

Dye layer CD recorders use a laser to discolor the dye layer in a CD-R, which allows it to scatter laser light in the same way as a pit on a commercially made CD-ROM.

E

EIDE Short for *Enhanced Integrated Drive Electronics*. The most popular interface standard for connecting internal hard drives and CD/DVD drives to your PC. A typical PC has both primary and secondary EIDE connectors; each connector can handle two EIDE devices (for a total of four drives).

El Torito format A format standard used to record bootable CD-ROM discs for use in PCs.

External drive A device such as a hard drive or DVD recorder that is placed next to your computer (instead of inside the case). External drives are connected to your computer by a cable and usually need their own power supplies.

F

FireWire A type of connection that allows external devices such as recorders, DV camcorders, and scanners to exchange data with your computer at a very fast rate. FireWire connections also allow you to control a DV camcorder from your computer.

FireWire is the common name for the IEEE-1394 standard, which was originally developed by Apple.

Formatting The process of preparing a hard drive, floppy disk, CD-RW, DVD-RAM, or DVD+RW/DVD-RW disc to store data. UDF/packet-writing programs such as HP DLA and Adaptec's DirectCD also require you to format a disc before you can use it.

G

Gigabyte (GB) A unit of data equal to 1,024 MB (megabytes).

H

Hard drive An internal or external device that stores data and programs, allowing you to save, move, and delete files.

HTML Short for *Hypertext Markup Language*. The programming language used to create pages for the World Wide Web. HTML can also be used to create menus for CD and DVD discs.

HTML Editor A text-based program used to create or edit HTML pages.

I

Incremental multisession disc A multisession disc with data imported from a previous session. Incremental multisession discs allow you to update existing data on a disc that you've already recorded.

Interface A standard method of connecting a hardware device to your computer. For example, DVD and CD recorders typically use EIDE, FireWire, and USB interfaces.

Internal drive A hardware device that fits inside your computer's case—for example, a hard drive, floppy drive, or DVD recorder.

ISO 9660 The original CD-ROM file system standard that is supported by virtually all computer operating systems. ISO 9660 is a good choice when recording a CD-ROM that will be read on many different types of computers, but it doesn't support long filenames.

J

Jewel box A plastic storage case for a compact disc. Jewel boxes can contain a cover and back insert. DVDs should not be stored in a standard CD jewel box; always use a DVD storage box instead.

Joliet file system The CD-ROM file system developed by Microsoft for Windows 95. Joliet allows long filenames of up to 64 characters that can include multiple periods and spaces.

Jumper A small wire and plastic electrical crossover designed to connect two pins on a computer circuit board. Most EIDE drives use a jumper to set the drive's master/ slave configuration during installation.

K

Kilobyte (KB) A unit of data equal to 1,024 bytes.

L

Land An area on the surface of a compact disc that reflects light. Recorded CDs and DVDs use a clear area within the dye or crystalline layer to act as a land.

M

Master A configuration setting used with EIDE drives to specify that the drive is the primary device on the EIDE cable. Your recorder should be set to "single drive, master unit" (if it's the only drive on the cable) or "multiple drive, master unit" (if it's the primary drive and there is another EIDE device on the same cable).

Megabyte (MB) A unit of data equal to 1,024 KB.

Mixed-mode disc A CD recording format that combines both digital audio tracks and a data track on one disc. Because the first track is recorded as computer data, a mixed-mode disc can't be played on an audio CD player.

Motherboard Your computer's main circuit board—it holds the processor and RAM chips, and most of the circuitry. You can expand your computer's features by plugging adapter cards into your motherboard.

MP3 The most common and popular digital audio format for storing music on your hard drive (or as files on a CD or DVD). With a recorder, you can burn audio CDs with CD-quality stereo MP3 files.

MPEG Short for *Moving Pictures Expert Group*. The video format used on commercial DVD movie discs. MPEG video can also be recorded to Video CDs.

Multisession A multisession disc (sometimes called *CD-ROM XA*) can carry separate discrete recording sessions, each of which can be accessed one at a time. There are two different types of multisession discs: incremental (where only the last session is available) and multivolume (where each of the sessions can be read). Older CD-ROM drives may not be able to read multisession discs.

Multitasking Running more than one program or application on your computer at once.

Multivolume multisession disc A multisession format that stores data in separate volumes on a single disc, each of which can be accessed one at a time.

N

Network A connected group of computers that can share files, printers, recorders, and modems through a cable or wireless system. Recording over a network is usually not a good idea.

P

Packet-writing (Also called *UDF recording*) A method of recording data directly to a disc, without running special recording software. A program such as HP DLA allows you to copy files directly to the recorder, just as you would with a hard drive.

Pit An area on the surface of a compact disc that scatters light (without reflecting it). Recorded CDs and DVDs use an opaque area within the dye or crystalline layer to represent a pit.

R

RAM Short for *random access memory*. Your computer's RAM temporarily stores programs and data while your computer needs them; when the computer is turned off, the information stored in RAM is lost.

Red Book The common name for the international standard that specifies the recording layout of audio CDs. In order to be read by an audio CD player, an audio disc must conform to the Red Book standard.

S

Safe zone The viewable area of a digital video clip when it's displayed on a television set.

Screen printing A method of labeling CD and DVD discs involving the application of layers of different inks in stages through stencils; the end result is a single multicolored image.

Secure connection An encrypted connection to a Web site that protects your personal data from a hacker; never use an online Web store unless it offers a secure connection.

Slave A configuration setting used with EIDE drives to specify that the drive is the secondary device on the EIDE cable. A DVD recorder can be set to "multiple drive, slave unit" if it's the secondary drive and another EIDE device acts as the primary master device on the same cable.

Slideshow A display program that shows images and video on your computer's monitor.

Static electricity Residual electricity that can damage your computer's hardware. Because static electricity can be stored in the body, never touch your computer's motherboard or adapter cards without first touching the metal chassis of your computer.

T

Tag A separate line in an HTML file that determines the display or specifies the contents of the page.

Thumbnail A miniature representation of a much larger image—a display of thumbnail images makes it easy to search for a specific digital photograph on a disc or your hard drive.

Timeline A control used in video editing that allows you to "assemble" and synchronize the image, soundtrack, and any transitions or effects into a cohesive whole.

Track On an audio CD, a track is a single section of audio (typically, a single song) that you can immediately access—data CDs also store information in one or more tracks.

Track-at-once A disc recording mode where the laser writes each data or audio track individually. The laser is turned off between tracks.

Transition An optical effect placed between two video clips to act as a segue; common effects include dissolves, fades, and wipes.

U

UDF Short for *Universal Disc Format* (also called *packet-writing*). A method of recording data directly to a disc, without running special recording software. A program such as HP DLA allows you to copy files directly to the recorder, just as you would with a hard drive.

UPC Short for *Universal Product Code*. A unique number assigned to every commercial audio CD. Most audio CD player programs and audio CD cataloging programs can use the UPC number to identify the disc and download the artist, title, and track information, if necessary.

USB Short for *Universal Serial Bus*. A connection for external devices such as hard drives, scanners, digital cameras, and CD/DVD recorders on both the PC and the Macintosh. USB devices can be plugged in and used without rebooting your computer.

V

Video CD A recording format that allows you to store MPEG digital video with a simple menu for viewing on a DVD player or video CD player.

Virtual memory An increase in the amount of RAM available to your computer, accomplished by your operating system temporarily using hard drive space to store additional data. With virtual memory, Windows can run applications that require more RAM than you physically have available in your computer.

W

WAV Microsoft's standard format for digital sound files. CD-quality WAV files can be recorded in Red Book audio CD format, using most CD recording software.

Index

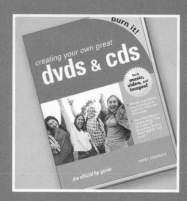

creating your own great dvds & cds

mark l. chambers

© 2003, paper, 368 pages,
0-13-100105-1

Make the most of any DVD or
CD recorder with this book.
Through start-to-finish projects,
you'll learn to create every type
of DVD, CD, and VCD, and record
whatever you want — video,
photographs, music, or data.
Coverage includes choosing the
right recorder, installation, formats,
software tools, slide shows, labels,
on-disc menus, troubleshooting —
even the latest DVD writers!

upgrading your hp pavilion pc

tom sheldon

© 2003, paper, 736 pages,
0-13-100415-8

Make your desktop HP Pavilion
PC better, faster, and more useful!
Whatever HP Pavilion you own —
old or new — this HP authorized
guide shows you exactly how to
supercharge it! One step at a time,
long-time PC expert Tom Sheldon
shows you what to buy, what to
do, and exactly how to do it. It's
cheaper than you think — and
easier than you ever imagined!

hp pavilion pcs made easy

nancy stevenson

© 2003, paper, 416 pages,
0-13-100251-1

Make the most of your new deskto
HP Pavilion PC and Windows XP
This book covers all you need to
get productive with your desktop
Pavilion fast — and have more fu
too! This book is a complete, eas
introduction to your Pavilion com
puter and Microsoft's Windows
XP. Easily master XP's amazing
collection of tools and learn to
configure XP exactly the way you
want it. Sure, there are other
introductory PC books — but this
is the only one that's authorized by
HP, and written just for you, the
Pavilion owner!